AN INTRODUCTION
TO AQUATIC TOXICOLOGY

AN INTRODUCTION TO AQUATIC TOXICOLOGY

MIKKO NIKINMAA

Professor of Zoology,
Department of Biology, Laboratory of Animal Physiology,
University of Turku, Turku, Finland

ELSEVIER

AMSTERDAM • BOSTON • HEIDELBERG • LONDON
NEW YORK • OXFORD • PARIS • SAN DIEGO
SAN FRANCISCO • SINGAPORE • SYDNEY • TOKYO

Academic press is an imprint of Elsevier

Academic Press is an imprint of Elsevier

The Boulevard, Langford Lane, Kidlington, Oxford, OX5 1GB, UK
225 Wyman Street, Waltham, MA 02451, USA

British Library Cataloguing in Publication Data
A catalogue record for this book is available from the British Library

Library of Congress Cataloguing in Publication Data
A catalogue record for this book is available from the Library of Congress

ISBN: 978-0-12-411574-3

For information on all Academic Press publications
visit our website at store.elsevier.com

15 16 17 18   10 9 8 7 6 5 4 3 2 1

Working together
to grow libraries in
developing countries

www.elsevier.com • www.bookaid.org

Contents

18. Modeling Toxicity

Preface

Any effects of contaminants on ecosystems must be caused by their direct effects on one or more organisms in the ecosystem. Thus, changes in the function (= physiology) of an organism or organisms will be behind any ecotoxicological effects. Consequently, understanding the normal physiology of organisms is at the center of all toxicology. Effluent discharges (and thus the measurements of chemical concentrations) are of no toxicological significance if they cause no effects on any organism. Further, a contaminant may cause effects on the genome of an organism without phenotypic effects. In such a case, the effect on the genome is not toxicological. Even though this premise emphasizes the role of functional changes in individual organisms, it is pointed out that interactions between organisms, and between natural environmental and toxicant-induced changes, occur whenever toxicant responses of organisms in an ecosystem are considered. For example, it is emphasized that many of the effects of toxicants in an ecosystem are indirect: the actual toxic effect is on a species other than the one that is seen to respond. The response takes place because of changes in biotic interactions within or between species. Further, while experimental studies are mainly retrospective, modeling makes aquatic toxicological predictions possible. Consequently, aquatic toxicology is truly integrative, and I hope that I am able to give this message to readers of the book. Since the book is introductory, I have tried to find a balance between making it possible for any reader to understand the book and including new concepts to make it interesting reading, even for aquatic toxicology experts.

Because my background is in animal physiology, the book is biased towards the animal kingdom. However, I hope that it becomes clear to the reader that effects on prokaryotes, fungi, plants, and algae are equally important components of aquatic toxicology. Which group of organisms receives most emphasis depends mainly on the interests of the writer. Regardless of the group that receives the most attention, the basic principles of aquatic toxicology are the same. However, as my approach to toxicology is biological, the work may be difficult for any reader without previous biological knowledge.

Acknowledgments

My thanks go first and especially to my wife, for all the help with practicalities when I have been consumed with different aspects of writing. Second, all the students who have taken my ecotoxicology classes have contributed to understanding which aspects of aquatic toxicology are most interesting and which most difficult. Third, my research group and all the other people working in the Animal Physiology Laboratory of the Department of Biology at the University of Turku deserve thanks for making the day-to-day working environment good and fruitful for a huge project, as the writing of this book turned out to be. I also want to thank the Elsevier production team, and especially Molly McLaughlin, for helping to fulfill the project. As a separate note, I was pleasantly surprised when Rhys Griffiths from Elsevier contacted me suggesting that I write the book. Finally, and perhaps most importantly, my sincere thanks go to the aquatic toxicologists whose thousands of manuscripts I have been privileged to see. Regardless of their final fate, they have made it possible to evaluate which points need to be emphasized in an introductory book in this field.

Every chapter ends with a very limited selection of relevant literature. I have included a reference only in one chapter, although it is clear that several references have material and conclusions that would also be relevant for other chapters. The literature selection is by no means exhaustive, and undoubtedly does not even include all the most important references on the topic. I apologize for any significant omissions, but hope that the selected articles and books, as well as their literature lists, help readers when they are pursuing any topic further. The text also contains a glossary, again not inclusive, of terms that aquatic toxicology students frequently come across. The figures have been drawn using SmartDraw 7, SigmaPlot 12.3, and PowerPoint 2010 software, frequently using Motifolio templates. Hopefully the figures will clarify the text.

1

Introduction: What is Aquatic Toxicology?

Abstract

The chapter first discusses the history of aquatic contamination, highlighting major cases where aquatic contamination has become an issue and cases where efficient solutions to environmental problems have been reached. Thereafter, the hierarchy of biological functions that can be disturbed by toxicants is briefly introduced. Notably, even when the ultimate goal of toxicological research is to find out how contamination affects an ecosystem, one must remember that the toxicants affect molecular functions of the most sensitive species. Toxicological testing and its uses are then introduced. The principal available aquatic toxicology testing methods, as given by the International Organization for Standardization (ISO) and the Organisation for Economic Co-operation and Development (OECD), are tabulated and the procedure for validating toxicity tests internationally is given. An overview of important issues for aquatic toxicological research in the future is also given.

Keywords: acid rain; paper- and pulp-mill effluent; toxicity test; gene expression; DNA methylation; epigenetics; direct effect; indirect effect; partial-life-cycle test; early-life-stage test; sediment toxicology; oil pollution; sublethal effects.

1.1 THE HISTORY OF AQUATIC TOXICOLOGY

As long as human populations remained small, anthropogenic influences on aquatic environments remained small and local. Any effects were in the vicinity of bigger settlements.

Probably the first larger-scale aquatic environmental issue resulted from lead water pipes that were used in large Roman towns.

Later, a major aquatic environmental problem was generated when sewage systems were built and people started using toilets (WCs). Consequently, contaminated household water, urine, and feces were disposed of directly to surrounding waters. Although cleaning measures are nowadays taken for most large human settlements, at least in Europe, Japan, North America, and Australia, the eutrophication caused by fertilizing compounds from human settlements, industry, agriculture (including the production of livestock), and aquaculture is a major threat to inland and coastal waters. Because gut bacteria can cause epidemics of intestinal diseases (e.g. cholera), they are still a major component to be determined when water quality criteria are established. The water quality framework is defined for Europe in the Water Policy Framework Directive (WFD) of 26 February 1997, and for the USA in the Clean Water Act and the Water Quality Act, of which the latter is from 1987.

Upon industrialization, acid rain became an issue. By the end of 1800s, coal burning was already causing acid rain and consecutive acidification of poorly buffered rivers and lakes in the British Isles. The immediate solution was to increase the height of chimneys. In the latter part of the twentieth century, this caused oxides of sulfur and nitrogen to be transported from central Europe and Britain to Scandinavia. The acid rain generated came down into poorly buffered streams and lakes in Norway, Sweden, and Finland, where whole fish stocks, especially of salmonids, were wiped out (see Figure 1.1 for the mechanism of water

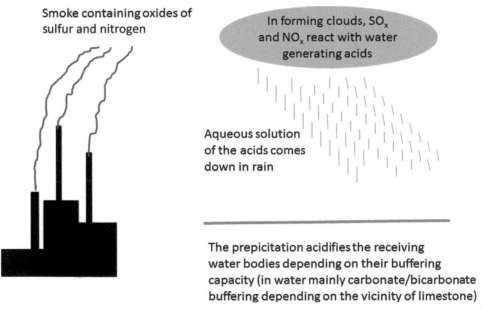

Smoke containing oxides of sulfur and nitrogen

In forming clouds, SO_x and NO_x react with water generating acids

Aqueous solution of the acids comes down in rain

The prepicitation acidifies the receiving water bodies depending on their buffering capacity (in water mainly carbonate/bicarbonate buffering depending on the vicinity of limestone)

FIGURE 1.1 **Schematic representation of the formation of acid rain.** The smoke contains oxides of sulfur and nitrogen (SO_x and NO_x), which react with atmospheric water to form H_2SO_3, H_2SO_4, and HNO_3. These acids are a part of precipitation and acidify waterways.

acidification). Similar acidification was observed in some parts of Canada, where it was caused mainly by coal burning in the industrial areas of the USA. As clear environmental disturbances were observed and could be tied to specific polluting sources, in this case sulfur-containing coal (and oil), and as the problems were observed in the aquatic systems of democratic industrialized European and North American countries, it was soon required that, first, the use of fuels containing much sulfur be curbed, second, the use of coal in energy production be decreased, and, third, the smoke be cleaned, removing sulfur from the gases. As a result of these measures, acid rain as an environmental problem is now all but forgotten in Europe and North America. Healthy fish stocks have returned to many formerly acidified lakes. However, globally, acidification of freshwater is of major importance, especially in Asia, where none of the measures that are required in Europe to prevent pollution are so far applied.

Until the latter part of the twentieth century, wastewater was virtually always uncleaned. Whenever the vicinity of effluent pipes became fouled, the solution was to increase the length of the effluent pipe. It was customary to talk about "the self-cleaning capacity of waters." Because of the idea that effluents could be fed into surrounding waters without cleaning, many major catastrophes occurred. For example, the toxic effects of mercury were seen in the Minamata incident in Japan. Tens or even hundreds of people died of mercury intoxication in 1956, as untreated effluents from a chemical factory were discharged in a bay where local inhabitants took their household water and ate the fish. Although the acute catastrophe could be pinpointed to the single year, the mercury contamination of the bay occurred between 1932 and 1968, and up to the present, around 2000 people have died with mercury intoxication being at least partially responsible, and more than 10,000 people have received some kind of compensation for mercury-intoxication-caused damages. Uncleaned paper- and pulp-mill effluents used to be a major environmental question in western Europe and North America. In the 1960s, the paper- and pulp-mill industry of Sweden and Finland produced an amount of effluent corresponding to the effluent produced by 100,000,000 people. At that time, all the water areas close to the paper and pulp mills were dead. Also, as a result of effluent discharge, the persistent organic pollutant (POP) concentrations (including polychlorinated biphenyls, PCBs) were so high that the reproduction of, for example, seals was very markedly affected. Since then, advances in paper- and pulp-mill technology have enabled the industry to be much more environmentally friendly: the use of chlorine in bleaching has been virtually discontinued, and the mills reuse most water. Consequently, the areas earlier uninhabitable for fish now have successful populations, and the gray seal populations in the Baltic Sea, for example, have increased markedly (Figure 1.2).

As a general conclusion from the history of aquatic toxicology, one can say that solutions to environmental problems are possible, but remediation and prevention of future problems require financial commitment. Thus, we should be prepared to pay some extra cost for products that contribute minimally to the deterioration of the aquatic environment. The decisions of consumers can ultimately change the ways of production. The directors of Scandinavian paper- and pulp-mill companies said in the 1960s that cleaning the effluents would not be possible as it would unacceptably reduce profits. However, when paper consumers started demanding cleaner paper, and began to leave environmentally costly products on the shelf, measures for producing environmentally friendlier paper were soon established.

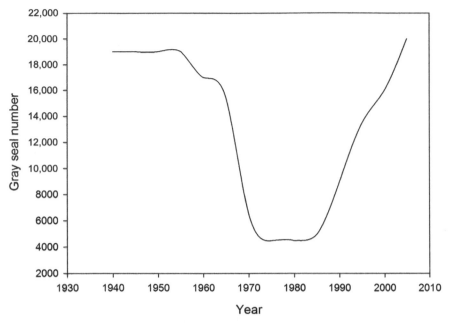

FIGURE 1.2 **Approximate population changes of the gray seal in the Baltic Sea from the 1940s to the 2000s.** Up to the 1940s, the population had decreased markedly from about 100,000 before the twentieth century because of intensive hunting. From the 1940s to the 1960s, the population remained stable, and then decreased in the 1960s and 1970s with the rise of a paper- and pulp-mill industry with poor effluent purification. The pollution of the Baltic Sea was associated with marked reproductive problems in seals. With the increasing efficiency of effluent cleaning, seal reproduction has again improved, and the population size is close to the value seen in the 1940s and 1950s. *Source: Harding and Härkönen (1999) and Harding et al. (2007).*

1.2 THE MAIN PRESENT AND FUTURE CHALLENGES

Since energy production and cars and other motor vehicles have needed more and more fuel, the demand for oil has continuously increased. Consequently, oil pollution associated with its exploration, refining, and transport has become a major challenge to aquatic toxicology. The problem becomes even more pronounced as oil exploration occurs increasingly in aquatic areas, in the Arctic and in deeper water than earlier. This means that major emphasis needs to be given to interactions with oil contamination and the natural environmental variables temperature and pressure. Another serious problem is that many different chemicals have been dumped in various aquatic bodies. The exact chemicals and even the places where dumping has occurred are often unknown. Important questions pertaining to aquatic environments involve sediment toxicology: how is the toxicity of a chemical affected by its adherence to the bottom sediment, what is the bioavailability of toxicants in the sediment, and how do toxicants move between the sediment and water? Since toxicants can affect each other's effects, it will be increasingly important to characterize these "cocktail effects." On a global scale, and as European and North American water purification standards are not used in many areas, the employment of universal water standards should be a priority, and include the costs involved in water cleaning. Water-cleaning units will be required everywhere, both for preventing the spread of disease-causing microbes in man and for preventing the deterioration of the aquatic environment with losses of organisms. A present global concern is the spread of antibiotic-resistant bacteria. Their increase is largely due to not having adequate water-cleaning standards for sewage from medical factories, at the moment especially in India. Another important issue is to evaluate the interactions between natural environmental variations and environmental contamination. Understanding such interactions is important for evaluating the consequences of chemicalization, e.g. for climate change and ocean acidification. Also, while eutrophication is not, as strictly defined, a toxicological problem, it and algal blooms are usually considered together with toxicological problems. The overall effects of aquatic contamination are reflected in fisheries: together with overfishing, the contamination of water is of the greatest influence on fish stocks and fish diversity. An important research area is always the appearance of new materials in the environment. At present this includes, in particular, the environmental effects of nanomaterials. One problem associated with these materials is that the traditional methodologies of aquatic toxicology may not be very suitable for studying their effects. These are just a few examples and personal views of the important issues for aquatic toxicology. They clearly indicate that the field has immense possibilities, and global problems.

1.3 WHAT IS MEASURED

1.3.1 The Cascade from Subcellular Genomic Effects to Ecosystem Effects

Toxicological effects must be effects on an individual, unless they are indirect. With indirect effects, one considers those that affect an organism via a toxicological effect on another species. As an example, a herbicide influences the occurrence of aquatic plants. Because of their decreased occurrence, hiding places for an animal decrease, and although no effects by the chemical on the animal occur, the indirect effect of the herbicide increases the likelihood of the animal being preyed upon. Another indirect effect could be that an organism itself is not affected at all, but the abundance of its preferred prey is decreased, whereby the nutritional state of the organism is affected.

Direct effects must affect the function of an organism. They need not influence the mortality, but can affect, for example, swimming speed, vision, number of offspring produced, and such like, which may all influence the abundance of sexually mature organisms of the next generation. Notably, there cannot be ecological effects without some organism being directly affected. However, often the studied organisms are not the ones the toxicant affects in nature, but have been chosen because of their common use in toxicity testing, because they are highly visible in the affected environment, or because they are important food items. In these cases, organisms that are likely affected, but not conspicuous, may escape attention in the short term. Also, the traditional end point in evaluating toxicity is death (see 1.3.2 Toxicity Testing). The lethality end point is very seldom directly encountered in nature. Rather, what are seen are sublethal effects, and as a definition for a toxicant effect in an ecosystem one could take the effect of contamination on the abundance of organisms of the next generation. If the abundance is not affected, contamination has likely not had an effect.

Figure 1.3 gives an overview of the stages of biological organization that can be affected by toxicants and be measured. First, toxicants can affect the genome: they can cause mutations,

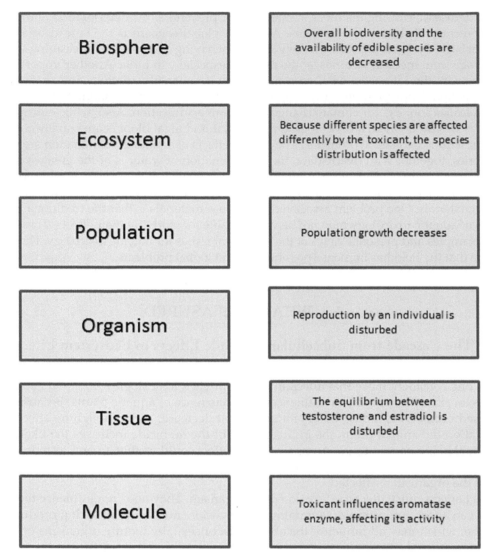

FIGURE 1.3   **The different levels of biological organization from molecular to global, with examples of how an environmental toxicant can affect the different levels.**

which are disadvantageous for the most part. However, occasionally beneficial mutations occur, which enable adaptation to a toxic insult. It is much more common that the genetic adaptation to toxic conditions is the result of some individuals carrying gene isoforms that enable more effective survival or greater offspring production than their conspecifics, which increases their proportion in a population. The genes are expressed mostly as the protein gene product. Thus, gene expression (Figure 1.4) has the following components: transcription, editing the transcript, transport of the transcript across the nuclear membrane, transformation of the transcript to mRNA and the stability of the mRNA, translation, folding of the amino acid chain,

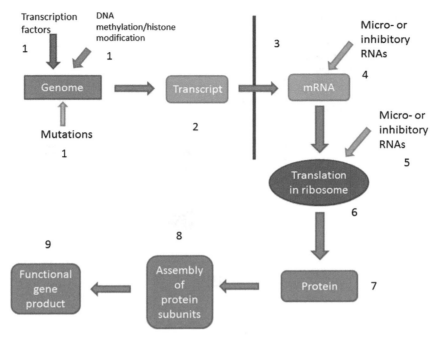

FIGURE 1.4 **The different steps of gene expression.** Gene expression consists of/is regulated by: (1) Transcription, which is measured by microarrays and quantitative polymerase chain reaction (PCR). This is often erroneously equated with gene expression. The efficiency of transcription can be altered by mutations in the transcribed gene (especially those occurring at the promoter/enhancer regions). (2) Editing of the transcript, e.g. its splicing to form multiple mRNAs encoding different splice variants of proteins. (3) Transcript transport across the nuclear membrane to form mature mRNA. (4) mRNA stability. (5) Translation of the mRNA to protein. Globally, translation, which is the most energy-requiring step of gene expression, is the step where gene expression is most importantly regulated. MicroRNAs and other regulatory RNAs control translational efficiency (and transcript splicing and mRNA stability). (6) Protein folding. (7) Protein stability. (8) Subunit assembly of multisubunit proteins. (9) Posttranslational processing of protein activity, e.g. phosphorylation. All the different levels of gene expression can be affected by environmental toxicants.

aggregation of multi-unit proteins, and the stability of the protein (including its breakdown to subunits). All the different steps of gene expression can be affected by environmental contamination. Also, as indicated in Figure 1.5, the amount or activity of the gene product (protein) can change differently from the transcript/mRNA level. Because of these points, equating gene expression to transcription is erroneous. Furthermore, since the protein gene products, which can change differently from transcriptional changes, carry out the various functions, it is also a mistake to say based on template (transcript/mRNA) changes that the mechanism of toxicant action would have been studied. As an analogy, saying on the basis of mRNA changes that a function has been affected is like saying that as we have printed more instruction books on how to build a machine, an increase in the product that the machine makes has increased. Studies on the mechanism of toxicant action require that it is known how the function of the gene product changes, and this requires measurements of the relevant protein activity. A feedback loop from protein activity to the genome results every once in a while in the following way: The protein activity decreases. Information about this is relayed to the genome. The transcription of

FIGURE 1.5 **An example of how environmental pollution affects the transcript levels and enzyme activity of (A) catalase and (B) glutathione peroxidase.** For both transcript abundance and enzyme activity, the control value (black bar) is given the value of 100%. *Source: Giuliani et al. (2013), on the mussel Mytilus galloprovincialis.*

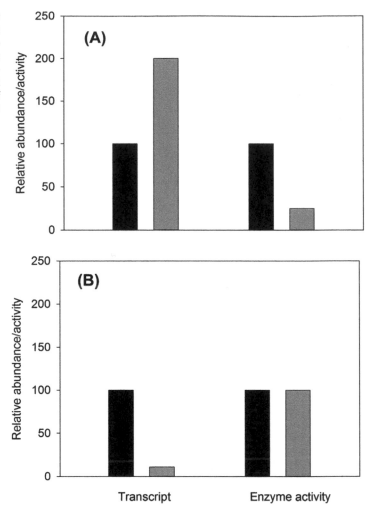

the relevant gene increases to increase protein production to "correct" for the reduced protein activity. Now, if mRNA change is equated to functional change, a completely fallacious conclusion is drawn. The lack of correspondence between transcriptional and protein-level responses is not just theoretical, but has been observed in many studies trying to explain the functional consequences of toxicant exposure. Further, in many cases, the transcriptional and functional responses and their correspondence to toxicant exposure differ depending on the duration of exposure, exposure concentration, and water quality during the exposure.

One way of regulating transcription is to influence histone modification or methylation of DNA. These affect chromatin coiling. The tightly coiled chromatin is usually transcriptionally inert. Upon methylation/acetylation changes (usually demethylation/deacetylation) the coiling is affected and transcription can proceed (Figure 1.6). One of the causes of both tissue-dependent and developmental-stage-dependent differences in gene transcription is different

FIGURE 1.6 **Inactive genes are often tightly coiled around histones in nucleosomes.** One possibility for gene activation in such cases is that changes in DNA methylation (normally demethylation) or histone modification open the coil and allow transcriptional activation. Such effects on DNA methylation/histone modification can be a reason behind tissue-specific, developmental-stage-specific, and epigenetic differences in gene expression, and toxicant effects on these. (1) Coiled DNA; (2) the coil is opened as a result of demethylation, allowing (3) the assembly of general transcription factors (A) on the uncoiled chromosome and consecutive transcription.

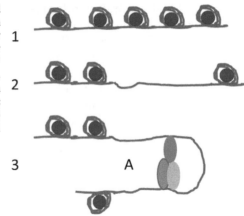

methylation/coiling of the chromatin, affecting how genes can be transcribed. Consequently, one important toxicological alteration, which will be different in different tissues and onto-genetic stages, is effects on DNA methylation and histone modification. These changes can also be behind epigenetic effects. Epigenetic effects are defined as transgenerational effects that are not caused by changes in gene structure. If a change in methylation/acetylation of histone or chromatin persists in generations following the one where it was first generated, an epigenetic effect is observed. Clear examples of aquatic toxicants affecting methylation have been observed, but good examples of transgenerational transfer of such effects are very few.

The overall functional responses, as seen in detoxification, excretion, energy production, movement, etc., are only seldom caused by the function of single genes. Rather, any observed effect is likely to be the result of changes in the expression of many genes. Importantly, the same net effect can result from different expression changes of the genes in different path-ways, which complicates interpreting any integrative changes on the basis of observations on a single gene. Interpretation becomes even more complex when populations, communi-ties, and ecosystems are considered from the ecotoxicological angle. The complexity is further increased when natural environmental variables such as temperature are included in expo-sures. As a result, the ease of interpretation of a toxicological result is inversely proportional to its ecological relevance.

1.3.2 Toxicity Testing

The aim of toxicity testing is to evaluate toxicity using standard methods. One of the targets of toxicity testing is to derive water quality criteria. In toxicity testing, both the test organ-ism and the test water used are normally standardized, i.e. they are the same regardless of where the testing is carried out, be it Tromsø in Norway or Sao Paulo in Brazil. Since toxicity testing aims at standardizing the methodology used, quite elaborate procedures are applied before an international test is deemed ready. As an example, the way by which International Organization for Standardization (ISO) tests are approved is given. The ISO is the largest developer of voluntary international standards in the world. When a need for a standard is established, a technical panel gathers and discusses the available data and methodology on the topic. Finally, a draft standard is generated. This draft is distributed among ISO members, who then use it and comment on it. Thereafter, the members will vote on the inclusion of the protocol as an ISO standard. If consensus is reached, the method is taken as an ISO standard. A lack of consensus results in further development of the standard; it comes back to the technical panel, which considers modifications suggested by the member states. This elaborate development of international toxicity test standards means two things: first, it takes quite a long time before a standard is accepted and, second, the end points used for toxicity testing are very conservative. In addition to ultimately becoming international ISO standards, the methods in development are often accepted as national standards. Because of the need to evaluate the international standard suggestions and the use of standardized national testing methods, there are national working groups for toxicity testing, commonly under ministries of the environment. Another aspect is that official toxicity testing can only be done by accredited laboratories. The accreditation is done by regional authorities and usually involves on-site visits, submission of written documentation of procedures used, and use of certified reference material for obtaining control values (for the tests). The accepted ISO toxicity test standards can be bought from http://www.iso.org, and a selection of them (either accepted or under development) is given in Table 1.1. Another major source of international toxicity test standards is the Organisation for Economic Co-operation and Development (OECD). The web pages of the OECD document library currently list more than 100 protocols for use (Table 1.2 gives a compilation of methods pertinent for aquatic toxicology). Unlike ISO standards, OECD standards are freely available. Also, the US Environmental Protection Agency (EPA) has an extensive compilation of aquatic toxicity testing methods, which can be found at http://water.epa.gov/scitech/methods/cwa/. Since toxicity tests aim at standardizing conditions, their ecological relevance can be questionable. The most common end point used for animals is lethality: in the natural environment, contaminant concentrations are seldom so high that they would cause direct lethality. Rather, sublethal effects on, for example, swimming and reproduction, avoidance of predators, or metabolic changes, are more common (notably, an increasing number of standard tests target sublethal endpoints; however, indirect effects are seldom addressed). Further, toxicity tests usually use water with specified properties, which may differ markedly from the water properties of the natural environment. Thus, before a study, it is quite important to consider whether one wants to evaluate toxicity in general or whether one is looking at what may take place in the specific environment.

The different types of aquatic toxicity testing are given in Table 1.3. The test types are commonly acute and chronic toxicity tests for a given life stage, or life-cycle (LC) tests, which consider the whole life cycle of an organism from birth to reproduction, and the survival and growth of the early life stages of offspring. The duration of the test is commonly 6–24 months with fish. Because of this, and the fact that complicated rearing facilities are needed for LC tests with fish, both partial-life-cycle (PLC) and early-life-stage (ELS) tests have been developed. In PLC tests, one normally follows the reproduction of adult organisms and the survival and growth of early life stages of the offspring. In ELS tests, one follows the embryo and its development and growth. The rationale for using PLC and ELS tests instead of full-life-cycle tests is that the early life stages are commonly the most toxicant-sensitive stages of animals, whereby tests including those stages approximate results obtained from full-life-cycle tests. Normally, work with invertebrates uses full-life-cycle tests only. This is because the studied organisms usually have short life cycles and can be maintained in small volumes

TABLE 1.1   A Selection of ISO Toxicity Test Standards for Aquatic Toxicity Testing (Either Approved or Under Development) as of 2013

Test Number	Test Title
ISO 6341:2012	Water quality—determination of the inhibition of the mobility of *Daphnia magna* Straus (Cladocera, Crustacea)—acute toxicity test
ISO 7346-1:1996	Water quality—determination of the acute lethal toxicity of substances to a freshwater fish (*Brachydanio rerio* Hamilton-Buchanan (Teleostei, Cyprinidae))—part 1: static method
ISO 7346-2:1996	Water quality—determination of the acute lethal toxicity of substances to a freshwater fish (*Brachydanio rerio* Hamilton-Buchanan (Teleostei, Cyprinidae))—part 2: semi-static method
ISO 7346-3:1996	Water quality—determination of the acute lethal toxicity of substances to a freshwater fish (*Brachydanio rerio* Hamilton-Buchanan (Teleostei, Cyprinidae))—part 3: flow-through method
ISO 8692:2012	Water quality—freshwater algal growth inhibition test with unicellular green algae
ISO 9509:2006	Water quality—toxicity test for assessing the inhibition of nitrification of activated sludge microorganisms
ISO 10229:1994	Water quality—determination of the prolonged toxicity of substances to freshwater fish—method for evaluating the effects of substances on the growth rate of rainbow trout (*Oncorhynchus mykiss* Walbaum (Teleostei, Salmonidae))
ISO 10253:2006	Water quality—marine algal growth inhibition test with *Skeletonema costatum* and *Phaeodactylum tricornutum*
ISO 10706:2000	Water quality—determination of long-term toxicity of substances to *Daphnia magna* Straus (Cladocera, Crustacea)
ISO 10710:2010	Water quality—growth inhibition test with the marine and brackish water macroalga *Ceramium tenuicorne*
ISO 10712:1995	Water quality—*Pseudomonas putida* growth inhibition test (*Pseudomonas* cell-multiplication inhibition test)
ISO 10872:2010	Water quality—determination of the toxic effect of sediment and soil samples on growth, fertility, and reproduction of *Caenorhabditis elegans* (Nematoda)
ISO/TR 11044:2008	Water quality—scientific and technical aspects of batch algae growth inhibition tests
ISO 11348-1:2007	Water quality—determination of the inhibitory effect of water samples on the light emission of *Vibrio fischeri* (luminescent bacteria test)—part 1: method using freshly prepared bacteria
ISO 11348-2:2007	Water quality—determination of the inhibitory effect of water samples on the light emission of *Vibrio fischeri* (luminescent bacteria test)—part 2: method using liquid-dried bacteria
ISO 11348-3:2007	Water quality—determination of the inhibitory effect of water samples on the light emission of *Vibrio fischeri* (luminescent bacteria test)—part 3: method using freeze-dried bacteria
ISO 11350:2012	Water quality—determination of the genotoxicity of water and wastewater—*Salmonella*/microsome fluctuation test (Ames fluctuation test)
ISO 12890:1999	Water quality—determination of toxicity to embryos and larvae of freshwater fish—semi-static method
ISO/DIS 13308	Water quality—toxicity test based on reproduction inhibition of the green macroalga *Ulva pertusa*
ISO 13641-1:2003	Water quality—determination of inhibition of gas production of anaerobic bacteria—part 1: general test

(Continued)

TABLE 1.1   A Selection of ISO Toxicity Test Standards for Aquatic Toxicity Testing (Either Approved or Under Development) as of 2013—cont'd

Test Number	Test Title
ISO 13641-2:2003	Water quality—determination of inhibition of gas production of anaerobic bacteria—part 2: test for low biomass concentrations
ISO 13829:2000	Water quality—determination of the genotoxicity of water and wastewater using the umu-test
ISO 14371:2012	Water quality—determination of freshwater sediment toxicity to *Heterocypris incongruens* (Crustacea, Ostracoda)
ISO 14380:2011	Water quality—determination of the acute toxicity to *Thamnocephalus platyurus* (Crustacea, Anostraca)
ISO 14442:2006	Water quality—guidelines for algal growth inhibition tests with poorly soluble materials, volatile compounds, metals, and wastewater
ISO 14669:1999	Water quality—determination of acute lethal toxicity to marine copepods (Copepoda, Crustacea)
ISO 15088:2007	Water quality—determination of the acute toxicity of wastewater to zebrafish eggs (*Danio rerio*)
ISO 15522:1999	Water quality—determination of the inhibitory effect of water constituents on the growth of activated sludge microorganisms
ISO/DIS 16191	Water quality—determination of the toxic effect of sediment and soil on the growth behavior of *Myriophyllum aquaticum*
ISO/DIS 16303	Water quality—determination of toxicity of freshwater sediments using *Hyalella azteca*
ISO 16712:2005	Water quality—determination of acute toxicity of marine or estuarine sediment to amphipods
ISO/CD 16778	Water quality—calanoid copepod development test with *Acartia tonsa*
ISO/CD 17244	Water quality—bioindicator of potential toxicity in aqueous media—determination of the potential toxicity of aqueous samples on bivalve embryo–larval development
ISO 20079:2005	Water quality—determination of the toxic effect of water constituents and wastewater on duckweed (*Lemna minor*)—duckweed growth inhibition test
ISO 20665:2008	Water quality—determination of chronic toxicity to *Ceriodaphnia dubia*
ISO 20666:2008	Water quality—determination of the chronic toxicity to *Brachionus calyciflorus* in 48 hours
ISO 21338:2010	Water quality—kinetic determination of the inhibitory effects of sediment, other solids, and colored samples on the light emission of *Vibrio fischeri* (kinetic luminescent bacteria test)
ISO 21427-1:2006	Water quality—evaluation of genotoxicity by measurement of the induction of micronuclei—part 1: evaluation of genotoxicity using amphibian larvae
ISO 21427-2:2006	Water quality—evaluation of genotoxicity by measurement of the induction of micronuclei—part 2: mixed population method using the cell line V79
ISO/TS 23893-2:2007	Water quality—biochemical and physiological measurements on fish—part 2: determination of ethoxyresorufin-O-deethylase (EROD)
ISO/FDIS 23893-3	Water quality—biochemical and physiological measurements on fish—part 3: determination of vitellogenin

TABLE 1.2   OECD Tests Relevant for Aquatic Toxicology

Test Number: Name	Description
236: Fish-embryo acute toxicity (FET) test	Intended for evaluating the acute toxicity of chemicals to zebrafish embryos: fertilized eggs are exposed to the test chemical for a period of 96 hours
210: Fish early-life-stage toxicity test	Test results define the lethal and sublethal effects of chemicals on the early-life stages of the species tested; the test is done using five concentrations of the test chemicals
211: *Daphnia magna* reproduction test	Assesses the effect of chemicals on the reproduction of *Daphnia* using young females
229: Fish short-term reproduction assay	Guideline describes an in vivo assay for fish reproduction; spawning-ready male and female fish are exposed to a chemical while being held together up to 21 days, i.e. a short part of the life cycle
201: Freshwater alga and cyanobacteria growth inhibition test	Determines the effects of a chemical on the growth of freshwater microalgae or cyanobacteria; the organisms are at the exponential growth phase
234: Fish sexual development test	Early-life-stage effects of a chemical are assessed with special emphasis on effects of putative endocrine-disrupting chemicals on the sexual development of fish
235: *Chironomus* sp. acute immobilization test	Guideline describes an acute immobilization assay on chironomids
209: Activated sludge, respiration inhibition test (carbon and ammonium oxidation)	Effects of a substance on microorganisms from activated sludge of wastewater-treatment plants are assessed by measuring their respiration rate (carbon and/or ammonium oxidation)
233: Sediment-water chironomid-life-cycle toxicity test using spiked water or spiked sediment	Effects of prolonged exposure of chemicals on the life cycle of sediment-dwelling *Chironomus* sp. are assessed using five concentrations of the test chemical
230: 21-day fish assay	Guideline describes an in vivo assay to study the effects of certain endocrine active substances on sexually mature male and spawning female fish, which are held together and exposed to a chemical during a limited part of their life cycle (21 days)
231: Amphibian metamorphosis assay	Screens substances that may interfere with the normal functioning of the hypothalamo–pituitary–thyroid axis with the help of amphibian metamorphosis
225: Sediment-water *Lumbriculus* toxicity test using spiked sediment	Evaluates how prolonged exposure to sediment-associated chemicals affects the reproduction and the biomass of the endobenthic oligochaete *Lumbriculus variegatus* Müller
224: Determination of the inhibition of the activity of anaerobic bacteria	Predicts the likely effect of a test substance on gas production in anaerobic digesters
221: *Lemna* sp. growth inhibition test	Evaluates the toxicity of substances to freshwater aquatic plants of the genus *Lemna* (duckweed) using exponentially growing plant cultures

(Continued)

TABLE 1.2   OECD Tests Relevant for Aquatic Toxicology—cont'd

Test Number: Name	Description
202: *Daphnia* sp. acute immobilization test	Evaluates the acute effects of chemicals on daphnids; usually the immobilization of daphnids aged less than 24 hours at the start of the test is followed
218: Sediment-water chironomid toxicity using spiked sediment	Effects of prolonged exposure to sediment-bound chemicals on sediment-dwelling larvae of freshwater *Chironomus* sp. are followed
219: Sediment-water chironomid toxicity using spiked water	Effects of prolonged exposure to chemicals in the aquatic phase on sediment-dwelling larvae of freshwater *Chironomus* sp. are followed
215: Fish, juvenile growth test	Evaluates how prolonged exposure to chemicals affects the growth of juvenile fish; normally fish in exponential growth phase are studied using five exposure concentrations
212: Fish, short-term toxicity test on embryo and sac-fry stages	Life stages from the newly fertilized egg to the end of the sac-fry stage are exposed to five concentrations of the test chemicals, and mortality (or other end points) followed
203: Fish, acute toxicity test	Fish are exposed to the test substance and mortalities recorded at 24, 48, 72, and 96 hours; LC50 is determined at each of these time points
210: Fish, early-life-stage toxicity test	Evaluates the lethal and sublethal effects of chemicals on the early life stages of the species tested using exposure to at least five concentrations of the tested chemical
204: Fish, prolonged toxicity test: 14-day study	Guideline gives advice for the determination of lethal and sublethal effects in fish exposed to test substances for 14 days; replaces test no. 203 if a longer observation period is considered appropriate
305: Bioaccumulation in fish: aqueous and dietary exposure	Procedure describes how to characterize the bioconcentration potential of substances in fish, using an aqueous (standard and minimized tests) or dietary exposure
302C: Inherent biodegradability: modified MITI test (II)	Describes the modified MITI test (II), which can determine e.g. the biochemical oxygen demand (BOD)
314: Simulation tests to assess the biodegradability of chemicals discharged in wastewater	Guideline describes how to assess the extent and kinetics of primary and ultimate biodegradation of organic chemicals that enter the environment initially in wastewater
315: Bioaccumulation in sediment-dwelling benthic oligochaetes	Allows evaluation of how sediment-associated chemicals are bioaccumulated in endobenthic oligochaete worms
316: Phototransformation of chemicals in water—direct photolysis	Designed as tiered approach to determine the potential effects of solar irradiation on chemicals in surface water; considers direct photolysis only
313: Estimation of emissions from preservative-treated wood to the environment	Emissions from wood and wooden commodities that are not covered and are in contact with freshwater or seawater can be estimated; preservative-treated wood test specimens are immersed in water

TABLE 1.2   OECD Tests Relevant for Aquatic Toxicology—cont'd

Test Number: Name	Description
311: Anaerobic biodegradability of organic compounds in digested sludge: by measurement of gas production	Guideline describes a method for the evaluation of anaerobic biodegradability
308: Aerobic and anaerobic transformation in aquatic sediment systems	Laboratory test method for assessing aerobic and anaerobic transformation of organic chemicals in aquatic sediment systems
303: Simulation test—aerobic sewage treatment—A: activated sludge units, B: biofilms	Two simulation tests used in aerobic sewage treatment, activated sludge units, and biofilms; designed to determine the biodegradation of water-soluble organic compounds in aerobic sewage treatment
305: Bioconcentration: flow-through fish test	Determines the bioconcentration potential of substances in fish under flow-through conditions
306: Biodegradability in seawater	Two methods for evaluating the biodegradability of a compound in seawater

The PDF protocols can be obtained from http://www.oecd.org/chemicalsafety/testing/oecdguidelinesforthetestingofchemicals. htm. The test number, title, and short description of the test are given.
LD50: lethal dose for 50% of organisms

TABLE 1.3   Important Types of Toxicity Testing

Acute toxicity tests	Usually evaluate lethality of the tested compound in the short term (24–96 h); instead of lethality as the end point, sublethal effects are increasingly used in testing
Chronic toxicity tests	The test is considered chronic if it encompasses more than 10% of the life-span of the organism; otherwise like acute toxicity tests
Life-cycle tests (LC)	Usually the lethal toxicity of a compound is followed throughout the development of an organism (from fertilization to sexual maturity); recently, sublethal end points have increasingly been used
Partial-life-cycle (PLC) tests	The toxicity evaluation includes life stages that are supposed to be most sensitive to chemicals
Early-life-stage (ELS) tests	Usually the embryotoxicity is determined; the rationale is that embryos and other early life stages are the most chemical-sensitive stages in the life of organisms
Reproductive toxicity tests	Determines reproductive end points (such as reproductive output) in acute or chronic toxicity tests
Developmental and embryotoxicity tests	Uses end points that are specifically associated with embryos/developing organisms; teratogenesis is one such end point
Frog teratogenicity assay (FETAX)	The test system determines the teratogenicity of chemicals on early frog (*Xenopus*) development; thus, the commercially available test actually belongs to developmental tests

(Continued)

TABLE 1.3   Important Types of Toxicity Testing—cont'd

Phototoxicity tests	Evaluate how the toxicity of chemicals is altered by sunlight
Behavioral tests	Determine how chemicals affect behavioral endpoints
Sediment toxicity tests	Investigate the toxicity of chemicals associated with sediment
Microtox	The method uses *Vibrio fischeri* bacterium, which emits light; light emission decreases with increasing toxicity of the sample (as the chemicals are toxic to the bacterium)
Reporter assays (e.g. CALUX)	Genetically modified bacteria/cells are exposed to the chemical; the genetic modification is done so that the chemical-responsive element drives luciferin gene expression, whereby emitted light intensity is directly proportional to exposure level

of water. The end points of toxicity testing commonly include lethality. Since lethality as an end point is only clear for vertebrates, for invertebrates and mobile microalgae, fungi, and prokaryotes, one often uses an immobilization assay instead, and growth inhibition tests for plants and prokaryotes. A common toxicity test with prokaryotes is Microtox. The testing method uses the luminescent marine bacterium *Vibrio fischeri* as the test organism, and the end point is the luminescence, which is affected by toxicants. Another newly developed set of testing methods uses genetically modified bacteria. Typically, a toxicant-responsive promoter is attached to the luciferin gene so that the luciferin bioluminescence changes as a response to the harmful compound. Hitherto, the methods have been available for testing estrogenicity and metal toxicity of water, but the methodology will certainly get wider use: theoretically, methods could be developed for all compounds with promoter sequences that increase transcription. The two major benefits for bacterial toxicity tests are that they are very rapid (with results normally obtained within an hour) and require a very small amount of water.

Relevant Literature and Cited References

Ankley, G.T., Schubauer-Berigan, M.K., 1995. Background and overview of current sediment toxicity identification evaluation procedures. J. Aquat. Ecosyst. Health 4, 133–149.

Blaise, C., Ferard, J.-F. (Eds.), 2005. Small-scale Freshwater Toxicity Investigations: Toxicity Test Methods. Springer, Berlin.

Bolsover, S.R., Shepherd, E.A., White, H.A., Hyams, J.S., 2011. Cell Biology: A Short Course, third ed. Wiley, Hoboken, New Jersey.

Boudou, A., Ribeyre, F., 1997. Aquatic ecotoxicology: From the ecosystem to the cellular and molecular levels. Environ. Health Perspect. 105 (Suppl. 1), 21–35.

Breitholz, M., Hill, C., Bengtsson, B.E., 2001. Toxic substances and reproductive disorders in Baltic fish and crustaceans. Ambio 30, 210–216.

Burton Jr., G.A., 1991. Assessing freshwater sediment toxicity. Environ. Toxicol. Chem. 10, 1585–1627.

Burton, G.A., Landrum, P.F., 2003. Toxicity of sediments. In: Middleton, G.V., Church, M.J., Gorgilo, M., Hardie, L.A., Longstaffe, F.J. (Eds.), Encyclopedia of Sediments and Sedimentary Rocks. Kluwer, Dordrecht, The Netherlands.

De Zwart, D., Slooff, W., 1983. The Microtox as an alternative assay in the acute toxicity assessment of water pollutants. Aquat. Toxicol. 4, 129–138.

Eto, K., 2000. Minimata disease. Neuropathology 20, S14–S19.

Giuliani, M.E., Benedetti, M., Arukwe, A., Regoli, F., 2013. Transcriptional and catalytic responses of antioxidant and biotransformation pathways in mussels, *Mytilus galloprovincialis*, exposed to chemical mixtures. Aquat. Toxicol., 120–127, 134–135.

Harding, K.C., Härkönen, T.J., 1999. Development in the Baltic grey seal (*Halichoerus grypus*) and ringed seal (*Phoca hispida*) populations during the 20th century. Ambio 28, 619–627.

Harding, K.C., Härkönen, T., Helander, B., Karlsson, O., 2007. Status of Baltic grey seals: Population assessment and extinction risk. NAMMCO Sci. Publ. 6, 33–56.

Helle, E., Olsson, M., Jensen, S., 1976. PCB levels correlated with pathological changes in seal uteri. Ambio 5, 261–263.

Kauppi, P., Anttila, P., Kenttämies, K. (Eds.), 1990. Acidification in Finland. Springer, Berlin.

Mothersill, C., Austin, B. (Eds.), 2003. In Vitro Methods in Aquatic Ecotoxicology. Springer, Berlin.

Reijnders, P.J.H., 1986. Reproductive failure in common seals feeding on fish from polluted coastal waters. Nature 324, 456–457.

Retief, F.P., Cilliers, L., 2006. Lead poisoning in ancient Rome. Acta Theologica. 26, 147–164.

Sahu, S.C. (Ed.), 2012. Toxicology and Epigenetics. Wiley, Hoboken, New Jersey.

Schindler, D.W., 1988. Effects of acid rain on freshwater ecosystems. Science 239, 149–157.

Stanbridge, H.H., 1976. History of Sewage Treatment in Britain. Institute of Water Pollution Control, Maidstone, UK.

Thompson, K.C., Wadhia, K., Loibner, A.P. (Eds.), 2005. Environmental Toxicity Testing. Blackwell, Oxford, UK.

US Environmental Protection Agency, Office of Water, 2002. Methods for Measuring the Acute Toxicity of Effluents and Receiving Waters to Freshwater and Marine Organisms, fifth ed. EPA-821-R-02–012.

US Environmental Protection Agency, Office of Water, 2002. Short-Term Methods for Estimating the Chronic Toxicity of Effluents and Receiving Water to Freshwater Organisms, fourth ed. EPA-821-R-02–013.

US Environmental Protection Agency, Office of Water, 2003. Short-Term Methods For Estimating the Chronic Toxicity of Effluents and Receiving Water to Marine and Estuarine Organisms, third ed. EPA-821-R-02–014.

What Causes Aquatic Contamination?

Abstract

The chapter details the most common aquatic contaminants. In addition to listing contaminants, the reader is first reminded that contaminants affect both climate change and ocean acidification by affecting the efficiency of photosynthesis of oceanic microalgae, which carry out almost half of global photosynthesis. Thereafter, it is pointed out that the toxicity of a compound is related not only to its amount in the environment but also to its water and lipid solubility affecting the uptake route, the fugacity of the compound (i.e. its tendency to dissociate from the matter it is currently associated with and associate with other constituents of the environment), its transformation—occasionally the transformed products are more toxic than the parent compound, and finally the complex formation, which often reduces the toxicity of a compound. A listing of toxicants is then made according to toxicant type, with an emphasis on which is the main source or use of the toxicant class. Thus, pesticides, pharmaceuticals, and paper- and pulp-mill effluents, for example, are treated each as one entity, although each group has several types of compounds having different modes of action. Further, contamination is understood broadly encompassing both radiation and genetically modified organisms. The problems associated with defining the toxicity of nanomaterials are highlighted. Chemical contaminants are broadly divided into inorganic and organic categories, with metals being the major inorganic contaminants. The chapter then considers the major uses of the toxicants and their routes to the environment. The reasons why the contaminants may affect organisms are also briefly introduced.

Keywords: contaminant; metal; pesticide; nanomaterial; radiation; genetic modification; persistent organic pollutants; POP; paper- and pulp-mill effluent; halogenated hydrocarbons; dioxin; PAH; organometallic compounds; oil; pharmaceuticals and personal care products; eutrophication; ionic liquids.

2.1 INTRODUCTION

A general understanding that human actions cause deterioration of the aquatic environment was only reached about 50 years ago, and even now the environmental aspects of these actions are often forgotten when environmental effects and short-term economic gains are in conflict. Further, it appears that, at any given moment, only one environmental problem reaches the headlines. At present, such a problem is climate change. In this context, one should remember that climate change and aquatic contamination are intimately intertwined. About 50% of all photosynthesis, which removes carbon dioxide, is carried out by photosynthetic aquatic organisms. Aquatic pollution has decreased the efficiency of photosynthesis, whereby toxicant effects on the aquatic environment facilitate climate change and ocean acidification.

The following factors need to be borne in mind when one considers the importance of a compound as a pollutant:

1. The amount of the compound released. Naturally, the more a substance is released, the more harmful it is. Comparing the released amounts is highly important, as it is possible that a potentially very toxic compound is environmentally unimportant if it reaches the environment only in very small amounts. Conversely, a compound which is relatively nontoxic may become an important contaminant, if it reaches the environment in large amounts. Examples of relatively nontoxic compounds that have caused serious environmental problems in aquatic bodies are nitrates and phosphates. Their release results in eutrophication and consequent lack of oxygen.
2. The water solubility of the compound. This affects both the uptake route (water vs food) and the biomagnification of the compound in the food web. This is closely related to bioavailability (see Chapter 6).
3. The fugacity of the compound. The tendency of the compound to escape from the material it is currently associated with will affect all aspects of its toxicity.
4. The transformation of the compound. Many chemicals that enter the environment do not remain in their original form, but are transformed into daughter compounds. The compounds formed may be more or less toxic than the parent compound. Both abiotic and biotic factors may cause the transformation. For example, solar radiation will cause oxidation of many compounds. Other compounds may be sensitive to oxygen level or pH

in the medium. Several organic compounds are biotransformed in organisms by specific pathways.

5. Complex formation by the compounds. Some chemicals form complexes with other constituents of the aquatic environment or the internal milieu of the organism. This will affect both the availability of the compound to an organism (bioavailability, see Chapter 6) and the ability of the compound to interact with its toxicity target.

When thinking about toxicants, it is important to remember that all compounds can become toxic, if an organism gets too much of them. This is particularly well demonstrated with metals such as zinc and copper; they are important constituents of enzymes, and must be obtained in small amounts. However, high concentrations are toxic, and aquatic toxicity studies on zinc and copper are very common. It should also be noted that in the following, I have taken a broad definition of toxicant, including, in addition to toxic compounds, radiation and genetic modification, although neither is a narrowly defined toxicant. Table 2.1 gives a list of the major factors causing environmental problems in aquatic systems. They are discussed in more detail below. It should also be noted that, although the uses of many chemicals are in no way associated with the aquatic environment, they appear there because waste treatment has traditionally concentrated on removing waste out of sight in wastewater pipes.

TABLE 2.1   Major Types of Aquatic Contaminants

Metals and metalloids	Includes essential metals such as copper, zinc, and iron; nonessential metals such as cadmium, lead, mercury, and silver; and metalloids such as arsenic. Major sources are household effluents, mining, and associated industry (e.g. smelteries), fertilizers, fuels, and well water
Organometallic compounds	Contaminants include organic tin compounds and methylmercury. Although methylmercury is partially of anthropogenic origin, natural methylation/demethylation processes also cause its presence in waterways. Organic tin compounds used to be important components of antifouling paints of boats and ships but have now been banned
Fertilizers	Include especially nitrates, ammonium nitrogen, and phosphates. Their sources are household effluents, agriculture, and aquaculture
Greenhouse gases	Carbon dioxide production is involved in ocean acidification, and methane can be liberated in natural gas production
Oxides of sulfur and nitrogen	Their deposition in smoke from energy production and traffic causes acid rain
Radioactive compounds	A natural source of radioactivity is radon gas. In addition, effluent from plants performing military processing (enriching) of uranium is a major source of radioactivity in water. If nuclear power plants are functioning properly, the radioactivity given out to water is smaller than from power plants using coal. However, abnormal occurrences, such as earthquakes or accidents in the power plants, may result in high environmental radioactivity. Further, leaks from storing radioactive material and uranium mining can be significant sources of radioactivity in water

(Continued)

TABLE 2.1   Major Types of Aquatic Contaminants—cont'd

Oil and its components	Oil spills can take place during oil drilling, as a result of shipwrecks during oil transport, and from effluent discharges from oil refineries. Also, oil tanks of ships are surprisingly often cleaned in open sea, resulting in oil discharges. The most toxic oil components are compounds with aromatic rings
Pharmaceuticals and personal care products	Since the intention is to produce drugs with minimal breakdown, these often pass through water cleaning unmodified. In addition, antibiotics may kill bacteria in biological water purification. Soaps and other detergents dissolve lipid membranes, and sunscreens are often photochemically modified to more toxic compounds
Halogenated compounds	Persistent organic pollutants (POPs) include halogenated compounds as a major component. Important halogenated compounds are polychlorinated biphenyls (PCBs), dioxins (e.g. TCDD), furans, and organochlorine insecticides. The appearance of new chlorinated organic compounds in the environment has markedly decreased recently, for two reasons: first, the reliance of paper bleaching on chlorinated compounds has markedly decreased and, second, the use of organochlorine insecticides (such as DDT) is severely restricted. However, chlorinated organic compounds are still a group of chemicals of concern, as they are highly persistent and bioconcentrate. They are further present in sediments in the vicinity of paper mills. Brominated organic compounds are extensively used in flame retardants. Fluorinated compounds are also increasingly found in the environment
Paper- and pulp-mill effluents	Since chlorinated compounds have disappeared from effluents, the major toxic compounds are natural compounds of trees, such as resin acids from coniferous trees and phenolics from deciduous trees
Endocrine-disrupting compounds	These include several types of compounds with various modes of action. Although several different types of hormonal pathways could be targeted, the term is most commonly used for compounds that disturb reproductive hormone cycles
Pesticides	Pesticides contain several different types of compounds, including herbicides, insecticides, and fungicides
Nanomaterials	The definition of a nanomaterial is any material with a maximal dimension of 100 nm. The use of nanomaterials has increased markedly during recent years, and nanotoxicology has consequently gained importance. This requires development of new methods, as conventional methodology is poorly suited for determining nanomaterial toxicity
Ultraviolet (UV) radiation	The effects are restricted to surface layers of water bodies, as light penetrates only a short distance in water
Ionic liquids	These compounds are called "environmentally friendly" solvents, because their vapor pressure is small; however, their aquatic effects are poorly known
Genetic modification	Contaminant effects of genetic modification are largely caused by methodological aspects

Contamination is defined broadly to include, in addition to chemical contamination, also radioactivity, UV radiation, and genetically modified organisms.

2.2 METALS, METALLOIDS, AND ORGANOMETALLIC COMPOUNDS

Metals are of two types. The so-called essential metals (Figure 2.1) are needed in small amounts in enzymes, oxygen-carrying pigments, etc., but become toxic in high concentrations. The nonessential metals are not needed, and they are toxic immediately after a limiting concentration has been reached. It has been customary to talk about heavy metals when metal toxicity is evaluated. The International Union of Pure and Applied Chemists (IUPAC) has advised against the use of the phrase. Instead, one could use the phrase

FIGURE 2.1 **The influence of amount of metals obtained on the survival of organisms.** (A) Essential metal: initially survival increases with increasing metal amount, until a maximum is reached when enough is obtained to fulfill the metal requirements (of, e.g. enzymes). Thereafter, an increasing amount of metal is toxic. Examples of essential metals are iron, zinc, and copper. (B) Nonessential metal: as the organism does not require the metal ion, an increase in metal amount does not initially affect survival, until a toxic level is reached, when an increase in metal amount decreases survival. Examples of nonessential metals are cadmium, lead, and mercury.

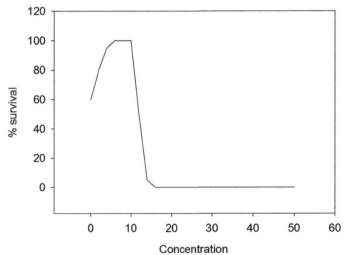

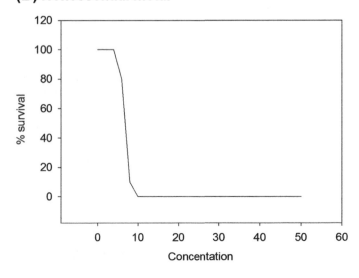

"metal contamination," and when required, define the metal accurately. Metals are present in the environment mostly as cations. The most common metals in organisms are sodium and potassium, which are needed in adequate amounts, for example, for normal nerve conduction. In addition to sodium and potassium, organisms need at least calcium, magnesium, iron, copper, and zinc. To obtain the needed metal ions, organisms have specific transporters and proteins that complex certain ions. The type of aquatic environment determines the environmental stresses associated with the major metal ions. In the case of freshwater, all organisms are hyperosmotic. In the marine environment, most organisms are isosmotic, with the exception that teleost fish are hyposmotic. Thus, in freshwater, organisms tend to lose salts and gain water, whereas marine teleosts tend to gain salts and lose water. This marked difference may have an effect on the mechanism or major site of toxicity of compounds that affect ion or water transport or permeability. With regard to the most common metals, sodium and potassium, they are seldom associated with toxic effects. An exception is the effect of road salts. Melting waters containing high salt concentration have caused erratic swimming in tadpoles. Road salting may also increase the salinity of groundwater. While an environmental excess of calcium is not known to cause toxicological problems, various environmental toxicants affect calcium metabolism in organisms, with consequences for cell signaling, bone or exoskeleton structure, and egg formation. Some metals exist at several valency states: these metals can participate in redox reactions, with the ion having the lower valency being more stable in reducing conditions (such as anoxia) and the ion with higher valency being stabilized in oxidizing conditions. The different ions have different toxicities, which indicates that conditions affecting metal speciation will influence metal toxicity. Metal speciation and the bioavailability of metals (see Chapter 6) are also markedly affected by complex formation both environmentally and within organisms. Metals reach the aqueous environment in mining and smeltery effluents, and in leachates from waste dumps. The specific sources of individual metals are described below, when the toxicologically important metals are introduced.

2.2.1 Copper

This metal is needed in small amounts, as its ion is a part of the active group of several enzymes and the invertebrate oxygen-binding protein hemocyanin. Copper is used especially in electric wires because it conducts electricity well. It is also used in water pipes, as it is easily malleable. The metal is also often a component of metal mixtures, and recently it has been used as a toxic component of boat paints. Of the two valency states of the ions of copper, the copper(I) (cuprous) ion is usually more toxic than the copper(II) (cupric) ion. The latter is stable in atmospheric conditions. The cuprous ion is stabilized in reducing (usually low-oxygen) conditions. The copper(II) ion is converted to the copper(I) ion in some tissues, e.g. before transport across cell membranes. Consequently, in some cases, the toxicity of copper is due to copper(I) even when the organism is exposed to copper(II).

2.2.2 Lead

Until recently, the organometallic compound methyl lead was one of the most important pollutants, as it was a component of leaded fuel. Leaded fuel is no longer used in industrialized countries. Lead from exhaust fumes caused mainly air pollution, so the aquatic pollution by the metal is largely from the metal being a component of paints, batteries, or piping. Thus, effluents of paint manufacturing, and industries and households using lead-containing paints or batteries are major aquatic pollution sources in addition to mining effluents. Another important aquatic pollution source has been lead in shotgun pellets, which has caused high lead levels in environments inhabited by waterfowl. Lead has also been used as weights for fishing nets, etc., because of its high density. Lead is a typical nonessential metal (Figure 2.1); hitherto no need for it has been demonstrated in organisms.

2.2.3 Cadmium

Cadmium is another toxicologically important metal with no known biological need. In nature it is always associated with zinc, and is consequently an impurity of all zinc-containing products, such as galvanized steel. It is also a common impurity in coal and phosphate fertilizers. Cadmium is used in metal alloys and in batteries. In addition to resembling zinc, cadmium often behaves like calcium.

2.2.4 Zinc

Organisms need zinc in the active groups of several enzymes. However, in high concentrations it becomes toxic causing, for example, lamellar thickening in gill epithelia of fish (and other gill-breathing animals). Consequently, oxygen uptake of water-breathing organisms can be impaired. A major use of zinc is in metal alloys (several steels) and in galvanization.

2.2.5 Iron

The ferrous (iron(II)) ion is a part of heme, i.e. the active group of all globins and cytochromes. Thus, a significant amount is needed. The absorption, transport, and storage of iron all require specific associated proteins. For example, iron is transported in the circulation bound to transferrin, and taken up into cells with the help of transferrin receptors. When these bind the iron-containing transferrin, they are taken up by the cells in clathrin-coated vesicles. These vesicles have a low pH value, and the iron is liberated from transferrin and largely taken up by another protein, ferritin. However, although iron is required, and elaborate mechanisms are used for its adequate acquisition, it is also a toxicant in high concentrations. For example, many small boreal lakes have a naturally high, toxic iron level. Iron can precipitate on gills as ferric oxide, impairing oxygen uptake. Iron (as also copper) can also undergo the Fenton reaction, whereby the very reactive hydroxyl free radical is formed.

2.2.6 Aluminum

Aluminum is one of the most common elements of the earth's crust. Its importance in toxicology has mostly been associated with environmental acidification. At high pH values (pH > 7), it forms mostly insoluble hydroxides; at intermediate pH (5–7), sparingly soluble hydroxides;

and at low pH values (pH < 5), the Al^{3+} ion. This ion is an osmoregulatory toxicant, whereas the hydroxides may precipitate on gills, hindering oxygen uptake. Thus, aluminum is considered to be an osmoregulatory toxicant at low pH values but a respiratory toxicant at intermediary pH values.

2.2.7 Silver

Silver was the most important effluent from traditional photography. The photographic images were formed as silver salts are highly light sensitive. In developing films, the excess of silver was removed and became part of effluent. With present-day digital photography, silver pollution has drastically decreased.

2.2.8 Arsenic

Arsenic is, in fact, metalloid. It is obtained as a byproduct of, in particular, copper mining. The uses of arsenic are mainly in pesticides and components of semiconductors. Lead batteries and lead pellets or bullets also contain some arsenic. It is well known to the general public, as arsenic has been used as a poison in many detective stories. Arsenic compounds were also the first chemical warfare agents, and the stockpiles of these have subsequently been dumped to the bottom of the sea. The aquatic environmental problem caused by arsenic is pronounced, especially in Bangladesh, where it is estimated that more than 50,000,000 people get their drinking water from contaminated wells. Many of the wells were dug in a United Nations program in the 1960s. In the short term, the intestinal diseases associated with microbiologically poor drinking-water quality decreased, but after a delay of a couple of decades, the harmful effects of arsenic started to appear; for example, the incidences of certain cancers increased markedly. Elevated arsenic concentrations are not limited to Bangladesh, but may occur anywhere in the world. High arsenic concentrations occur wherever water is in contact with iron-sulfide-containing sediments, which commonly contain high arsenic concentrations. Also, other metal sulfide ores are commonly associated with high arsenic levels. Thus, mining and metal-smeltery effluents (e.g. from gold, lead, copper, and nickel smelteries) contain harmful arsenic levels.

2.2.9 Mercury

It is difficult to draw the line between elemental mercury and organic mercury compounds; in particular methylmercury, as methylation–demethylation occurs naturally (Figure 2.2). Elemental mercury was used in thermometers, and is still used in batteries and light tubes (including so-called energy-saving light bulbs). Mercury is also used in extracting gold. For this reason, gold extraction has caused the death of fish in the Danube and the Amazon. Mercury contamination has reached the headlines; for example, because of the Japanese Minamata catastrophe (described in section 1.1). The catastrophe was largely caused because the major organomercurial, methylmercury, is lipid soluble and consequently bioaccumulates in organisms. The organometallic compounds are discussed more generally below. Notably, in Europe and North America, the amount of mercury reaching the environment has decreased by more than 90% from the maximal amounts of the 1960s.

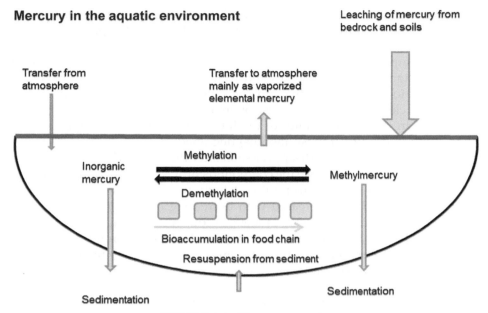

FIGURE 2.2   The mercury cycle.

2.2.10 Other Metals

Other toxicologically important metals are palladium and other palladium-group metals, because they are used in catalytic converters; rare earth metals, which are important in information-technology-product components; nickel, a common component of metal alloys (e.g. coins) and causing allergy in many people; cobalt, also often used in metal alloys; manganese, common in humic soils and possibly precipitating onto the gills of water-breathing animals in acid waters, when the pH of the gill microenvironment is higher than that of bulk water; tungsten; and uranium. Uranium as an environmental problem is usually associated with radioactivity, but the metal itself can be an environmental problem especially in the vicinity of open mines.

2.2.11 Organometallic Compounds

The major reason why organometallic compounds constitute an important environmental problem is that the organic group makes them lipophilic, whereby they usually bioaccumulate in organisms and can biomagnify (see Chapter 6 for a detailed discussion). The most significant compounds in aquatic environments are organic tin and mercury compounds. In addition, organic lead compounds were used in leaded fuel. Organic tin compounds, especially tributyl tin (TBT) but also triphenyl tin (TPT), were used as toxic components of antifouling paints of ship and boat hulls to prevent the growth of organisms such as, for example, algae and barnacles. As very different types of organisms must be affected, the primary mechanism of toxicity must be very nonspecific. One such mechanism could be

the following: TBT is a lipid-soluble weak acid; consequently, it dissipates proton gradients needed for mitochondrial energy production. If this were the primary mechanism of
toxicity, all aerobic organisms would be affected. A more specific effect led to the banning
of organic tin compounds in boat paints. They were observed to cause imposex in molluscs (female oysters developed a penis and were infertile). Also, fish from contaminated
areas were found to have reproductive problems. Because the concentrations present in the
environment had clear toxic effects, the use of organic tin compounds in ship paints is presently prohibited in the member states of the International Maritime Organization. As the
compounds are rapidly (in, at most, months) degraded in pure, oxygenated water, it was
thought that their toxicity would disappear soon after their ban, which had been enacted in
Europe by the 1990s. However, since the compounds proved to be more stable than expected
in natural environments, the reasons for this stability were examined. It was found that the
half-time of TBT degradation increased with decreasing oxygen level, such that it could
be up to several—even tens of—years in anoxic sediments (see Figure 2.3 for the half-life
of tributyl tin). The toxicity of organic tin compounds depends on the number of organic
groups present; apart from tetrasubstituted compounds, which are the least toxic, the toxicity decreases with decreasing number of organic groups. Thus, tributyl tin is more toxic
than dibutyl tin, which is more toxic than monobutyl tin. The major exception to this rule
is immunotoxicity, as described in section 11.9. The removal of organic groups is actually
an important detoxification mechanism of organic tin compounds. Organic mercury compounds, mainly methylmercury, are largely naturally occurring, although their levels can
increase anthropogenically. For example, phenylmercury was used to prevent the growth
of fungi in timber. Organic mercury compounds biomagnify in food webs, because they are
lipophilic. Also, as one goes to a higher trophic level, the proportion of organomercurials
increases and inorganic mercury decreases. About 90% of the mercury obtained by humans
is from fish, and the highest concentrations of mercury are found in top predators such as
pikes, and fish-eating birds and mammals. Since the sources of mercury are often natural,
e.g. soils and rock, high mercury levels are found in pikes and other predators from lakes in
uninhabited areas.

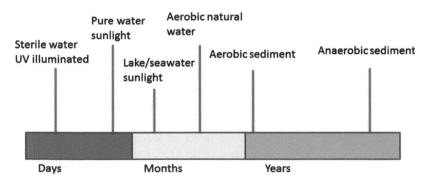

FIGURE 2.3 **The stability of tributyl tin chloride in different media.** The half-life of the compound increases
with decreasing radiation (UV light), and decreasing oxygen tension.

2.3 OTHER INORGANIC COMPOUNDS, INCLUDING FACTORS CAUSING EUTROPHICATION

In addition to metals, other inorganic contaminants in the aquatic environment are, first, nitrates and phosphates, which are nutrients and promote eutrophication. Second, hydrogen sulfide is formed and accumulated especially in anoxic bottom sediments. It is found in reducing environments, whereas reactive oxygen species (ROS) accumulate in hyperoxic conditions, which may occur in eutrophic waters during active photosynthesis. Third, the oxides of sulfur and nitrogen have been released in exhaust fumes and smoke, and have generated acid rain, which caused massive fish kills and disappearance of invertebrates with a calciferous exoskeleton from poorly buffered lakes, especially in Scandinavia and Canada (see section 1.1). Most smoke and exhaust fumes are now cleaned of sulfur oxides in Europe and North America. This has markedly decreased the acid rain problem in these areas. The remaining acidity is due to oxides of nitrogen, which cannot be removed as easily. Also, the acid rain problem has moved to Asia, where rapid economic growth has increased the use of fuels with high sulfur content, both in traffic and in energy production. Commonly, the exhaust fumes or smoke are little purified. Fourth, the increased anthropogenic production and decreased removal of carbon dioxide causes its accumulation and contributes to ocean acidification. Fifth, cyanide is commonly used in gold extraction, whereby it has caused massive fish kills in impacted rivers.

2.4 ORGANIC COMPOUNDS

Compounds containing carbon–carbon bonds are normally classified as organic, although some, e.g. halogenated ones, are not produced naturally. The compounds thus include alkanes and alkenes with and without branches, phenols, compounds with imidazole, and benzene rings. Because the range of compounds is vast, they are described below in groups according to their source, actions, or use. Toxicologically an important compound group are polyaromatic hydrocarbons (PAHs). They are not treated separately, but constitute an important toxic fraction of oil pollution, paper and pulp mill effluents, personal hygiene products, etc. Typically the lipophilicity of organic compounds is high, whereby they are taken up mainly in food, and may bioaccumulate and biomagnify in food webs. In aquatic environments, the hydrophobic compounds tend to distribute mainly to bottom sediment, which makes evaluation of their toxicity difficult. Also, because harmful compounds are stored in bottom sediments, their effects are seen long after new contamination has ceased.

2.4.1 Oil and Its Components

Oil drilling, transport, and use are all reasons for aquatic pollution. Major oil leaks have occurred because of, for example, the explosion at the *Deepwater Horizon* oil drilling platform off the coast of Louisiana in 2010, and as a result of big tanker wrecks (e.g. the *Prestige* in 2002 and the *Exxon Valdez* in 1989). However, although such major catastrophes make headlines, smaller leaks of oil occur daily: all oil production is associated with small leaks to the environment. As oil production increasingly takes place in shallow sea areas, with drilling depth increasing yearly, oil pollution at the production sites is increasing. This is associated with a drift of production sites to colder environments, increasing the need to study temperature effects on responses to oil. Shipwrecks releasing only a few hundred tons of oil in water occur regularly, and it is still surprisingly common to clean ship tanks of remaining oil, releasing the oil-containing cleaning water to the environment. Whenever oil leaks occur, pictures of oil-contaminated birds are shown. This is one of the problems, because when an oil-contaminated bird swims, water gets to the naked body surface. As the body temperature of birds is much higher than that of water, the energy consumption of the animals increases markedly as a result, and the energy production of the birds cannot keep up with requirements, ultimately leading to their death.

Oil does not consist of a single or a few compounds, but is a mixture of components with a varying number of carbon atoms per molecule (and varying toxicities). In oil refineries, the compounds are divided into size fractions and, thus, petrol, diesel oil, bitumen, etc. are obtained. Very light fractions constitute natural gas. Oil components are also the major building blocks of plastics. Oil is degraded by microorganisms, with decreasing efficiency as the temperature decreases. This fact is one of the reasons why Arctic oil exploration and drilling are particularly hazardous. Some of the compounds with one or several benzene rings (polyaromatic hydrocarbons; PAHs) are highly toxic. In combating oil spills, the oil slicks are often treated with compounds that dissolve them, making small oil droplets. The availability of toxic compounds from such droplets to aquatic organisms is much increased. The usually occurring concentrations of oil dispersants have not been shown to be acutely toxic, but dispersed oil is much more toxic to aquatic biota than the same amount not dispersed, obviously because of increased contact area with organisms.

2.4.2 Pesticides

Pesticides contain herbicides, fungicides, insecticides, molluscicides, and rodenticides. Table 2.2 gives an outline of major pesticides. A useful web page giving information on pesticides is http://www.pesticideinfo.org. In addition to their intended effects, concern needs to be attached to their effects on nontarget organisms. Some pesticides are much more toxic to nontarget organisms than to the ones that they are used against. Further, they often leach into water bodies, and thus their toxicity to aquatic organism forms a major concern. As an example, although atrazine formulations are targeted against terrestrial weeds, they appear to be quite toxic to amphibians, disturbing their early sexual development in environmentally relevant concentrations. Consequently, they are among the suspected reasons behind the recent amphibian declines. Many herbicides affect photosynthesis, and should thus have little direct toxicity to animals. However, as pointed out earlier, effects on plants can have indirect effects; e.g. on predator–prey relationships of animals, by affecting the availability of hiding places. The most commonly used herbicides at the moment are chloro-s-triazines, substituted phenylureas, and chloroacetanilides, specifically DCMU (3-(3,4-dichlorophenyl)-1,1-dimethylurea) (the most common formulation being Bauer's diuron) in Europe and glyphosate formulations in North America. The main target of insecticides is the transfer of neuroimpulses, especially synaptic transmission (see section 11.5). Although the use of insecticides is mainly terrestrial, they leach into water bodies, and are often more toxic to aquatic than terrestrial organisms. There is marked variation in the acute toxicity of insecticides to nontarget organisms (see Table 2.3). A problem with many insecticides is that the resistance of target organisms towards them develops rapidly, often in 10–20 years.

TABLE 2.2   The Major Pesticide Classes

INSECTICIDES	
Carbamates	The effect is the inhibition of cholinesterase, whereby nerve conduction and synapse function, especially at nerve–muscle junctions, is affected
Organophosphates	Examples are chlorpyrifos, malathione, and dimethoate. Organophosphates are also cholinesterase inhibitors. Because of their stability and effects on nontarget organisms, the use of many organophosphates has been restricted
Pyrethroids	These compounds are highly toxic to fish and other aquatic animals. They are commonly adsorbed to organic material of sediments
Organochlorines	This class of compounds is highly persistent, and bioaccumulate. DDT, aldrin, dieldrin, and heptachlor are compounds that have been banned worldwide, whereas e.g. lindane and methoxychlor are still commonly used. Organochlorine insecticides affect mainly synaptic transmission
Neonicotinoids	Chloronicotinyls and thionicotinyls are quite recent insecticide groups. They remain in plants for a long period of time after their concentration in soil has decreased below the measuring threshold
Microbe-derived insecticides	Most are compounds secreted by soil actinomycetes. Examples are spinosad and abamectin. The latter is highly toxic to aquatic animals
MOLLUSCICIDES (SEVERAL INSECTICIDES HAVE ALSO BEEN USED AS MOLLUSCICIDES)	
Metal salts	Although they are only weakly toxic, e.g. ferriphosphate and aluminum sulfate are used
Metaldehyde	Poisoning of nontarget animals is quite common
HERBICIDES (THE PURPOSE IS TO KILL WEEDS OR DEFOLIATE TREES)	
2,4-D (2,4-dichlorophenyxic acid)	The most widely used herbicide in the world. Synthetic compound with auxin-like action. Has dioxins as an impurity
Atrazine	Triazine herbicide that kills broadleaf weeds. Cheap, but quite stable. Affects photosystem II
Clopyralid	Used to remove broadleaf weeds from lawns. Synthetic auxin with stability as the major drawback
Dicamba	Used to remove broadleaf weeds from lawns. Synthetic auxin
Diuron (Bauerin brand chemically (3-(3,4-dichlorophenyl)-1,1-dimethylurea)	The most used herbicide in Europe. Inhibitor of photosynthesis
Fluroxipyr	A selective herbicide that is used to remove broadleaf weeds from crop fields; synthetic auxin
Glyphosate (e.g. Roundup)	Nonselective herbicide. Its use is based on simultaneous sale of genetically modified seeds that tolerate glyphosate. Glyphosate inhibits EPSP enzyme, which catalyzes a step in tryptophan, phenylalanine, and tyrosine production. Additional chemicals in the formulations may exert toxicity to aquatic organisms
Imazapic	A selective herbicide that is used especially in lawns
Linuron	Nonspecific inhibitor of photosynthesis

(Continued)

TABLE 2.2   The Major Pesticide Classes—cont'd

Metolaclor	Increasingly used instead of atrazine
Paraquat	More toxic to animals than any of the other commonly used herbicides. No longer in use in the European Union
Picloram	Used to prevent the growth of trees in meadows
FUNGICIDES	
Benzimidazoles	Inhibit mitotic division of fungal cells
Dithiocarbamates	Inhibit fungal growth
Famoxadones	Inhibit mitochondrial energy production
Fenamidones	Inhibit mitochondrial energy production
Chloronitriles	Inhibit fungal growth
Copper	Inhibits fungal growth
Sulfur	Inhibits fungal growth
Strobilurines	Inhibit mitochondrial energy production
Triazoles	Inhibit C14-demethylase, needed for sterol production

TABLE 2.3   An Example of How the Acute Lethal Toxicity of the Insecticide Malathion Varies Between Organism Groups

Organism	LC50 (μg/l)
AMPHIBIANS	
Frog (*Rana hexadactyla*)	1.24
Frog (*Rana tigrina*)	170
Frog (*Rana pipiens*)	2400
Frog (*Rana ridibunda*)	38,000
FISH	
Rainbow trout (*Oncorhynchus mykiss*)	120.8
Medaka (*Oryzias latipes*)	1493
Zebrafish (*Danio rerio*)	21,998
Ruffe (*Gymnocephalus cernua*)	96
CRUSTACEANS	
Aquatic sowbug (*Asellus aquaticus*)	60,620
Crayfish (*Procambarus clarkii*)	17,420

TABLE 2.3   An Example of How the Acute Lethal Toxicity of the Insecticide Malathion Varies Between Organism Groups—cont'd

Organism	LC50 (µg/l)
Crab (*Cancer magister*)	665.6
Northern pink shrimp (*Penaeus duorarum*)	12.2
Water flea (*Daphnia magna*)	10.2
Scud (*Gammarus fasciatus*)	1.12
MOLLUSCS	
Mussel (*Mytilus galloprovincialis*)	112,500
Great pond snail (*Lymnaea stagnalis*)	12,900
Unionid clam (*Indonaia caerulea*)	13

Data have been taken from http://www.pesticideinfo.org.

2.4.3 Endocrine-Disrupting Chemicals

Even though endocrine-disrupting chemicals should include all compounds affecting any hormonal pathway, the common use of the phrase is for compounds that disturb the normal reproductive hormone function (see section 11.4). Occasionally, disturbances of thyroid hormone function are also discussed. Because many different steps and pathways can be affected, the endocrine-disrupting chemicals comprise a wide variety of compounds.

2.4.4 Human and Veterinary Drugs, Personal Hygiene Compounds, and Detergents

Pharmaceuticals in the environment are of grave concern for several reasons. First, because drugs often need to be bactericides, they may kill bacteria in water-treatment plants where bacteria are used to remove organic contaminants (see Chapter 3 for the principles of water cleaning). Second, drugs tend to be stable and are poorly degraded. Consequently, they often pass through water treatment unchanged, especially if the water treatment is inefficient (or lacking). As a result, the environmental load of bactericides increases and antibiotic-resistant bacteria are generated. This is a problem especially in the drug-industry areas of India, where water cleaning is poor.

Soaps and household detergents contain compounds that decrease the surface tension of all lipids, thereby affecting the permeability properties of all organisms. This is particularly problematic for gill breathers, as an increase in passive gill permeability will easily disturb the ion and water balance. Among personal hygiene products, sunscreens deserve specific attention, as they are often compounds that are converted to more toxic ones by the ultraviolet (UV) radiation of the sun after entering the water. Notably, many personal hygiene products contain endocrine-disrupting compounds.

2.4.5 Paper- and Pulp-Mill Wastewater

The toxicity of effluents of paper and pulp mills built with northern European standards has decreased markedly because of decreased dependence on chlorine bleaching and more efficient cleaning of effluents. Also, modern paper and pulp mills typically reuse more than 90% of their water. Since chlorine bleaching is virtually completely abolished, the presence of, for example, polychlorinated biphenyls (PCBs) has decreased markedly in areas like the Baltic Sea. Also, the use of chlorophenolics, previously employed to prevent tainting of wood, has been discontinued, at least in North America and Europe. Because of these steps decreasing the toxicity of paper- and pulp-mill effluents, the most important toxic compounds are now wood-derived resin acids (from conifers) and phenols (from hardwood). Notably, at least in Scandinavia, the earlier sediments in the vicinity of the paper- and pulp-mill industry are more important sources of toxicants than the present effluents.

Effective water cleaning is not used in all paper and pulp mills throughout the world, so the effluent toxicity and the amount of little-treated effluent reaching the environment may differ markedly in different parts of the world. For example, as in Scandinavia in the 1960s, in China paper- and pulp-mill effluents often enter aquatic environments without being cleaned.

2.4.6 Halogenated Organic Compounds

Halogenated compounds are normally man-made. As a generalization, their toxicity increases with an increasing number of halogen moieties. The toxicity also increases with an increasing number of aromatic rings in the molecule. Previously, paper and pulp mills were the major source of halogenated compounds, especially chlorinated compounds because of chlorine bleaching, in Europe and North America. As chlorine bleaching has largely been substituted by ozonization, the amount of chlorinated hydrocarbons has markedly decreased, but the energy consumption of the paper industry has increased. A group of toxicants with significant importance are PCB compounds (polychlorinated biphenyls), comprising more than 200 congeners. In addition to being waste products of the paper- and pulp-mill industry, PCBs were used in lubricants and transformers. Although the compounds are now completely banned, and their amounts in effluents are very small, they are still compounds of concern, as they degrade very slowly and are found in sediments in the vicinity of paper and pulp mills. The toxicity of different PCBs varies markedly; because of this, their toxicity is often given as "dioxin equivalents," relating their toxicity to the corresponding toxicity of dioxin (TCDD). Some PCB congeners present virtually no dioxin-like toxicity. Thus, PCBs are occasionally divided into "dioxin-like" and "other" PCBs. Dioxin (or, actually, dioxins) is a family of halogenated compounds which are considered to be "supertoxins." However, their acute toxicity varies 10,000 times between rat strains. Rather than being acutely toxic, dioxins are teratogenic and tumor-causing, and exhibit developmental toxicity. Furans are closely related to dioxins. Neither group of chemicals was intentionally produced, but are either present as impurities of herbicides, PCBs, or phenolics, or generated during chlorine bleaching of paper and incomplete burning of chlorine-containing material.

The majority of chlorine-containing organic compounds have been banned because of their toxicity. However, several brominated compounds are still in use, especially as fire retardants (more than 70 different brominated compounds), although their toxicity can be expected to be similar to chlorinated compounds. Indeed, the oldest polybrominated chemicals have been proven toxic and their production and use have been prohibited—examples of such compounds are the polybrominated biphenyls (PBBs), which are a sister group to PCBs, and, recently, hexabromocyclododecane (HBCD). Also, some polybrominated diphenyl ethers (PBDEs), such as pentaBDE and octaBDE, are banned in some areas and generally classified among persistent organic pollutants (POPs). Altogether there are more than 200 possible congeners of PBDEs, with compounds containing a small number of bromine residues bioaccumulating most efficiently. The most commonly used fire retardant is decaBDE. The production and import of this compound group ceased in the USA at the end of 2013. Another much-used compound, especially in the epoxy resins of circuit boards, is tetrabromobisphenol A (TBBPA). Although this compound is highly toxic to aquatic animals (it is an endocrine disruptor; one reason for this being its resemblance to thyroxin), its environmental effects are thought to be small, as it is not thought to leak to the environment in significant amounts.

Fluorinated organic compounds (e.g. perfluorooctanoic acid) were originally thought to be quite inert and not reach the environment in significant amounts. The compounds are used, for example, in water-protective clothing. Recent studies have indicated significant distribution of the compounds in the environment, and as they are persistent, they may bioaccumulate to a significant degree.

2.4.7 Ionic Liquids

Ionic liquids consist of bulky organic cations (e.g. ammonium, imidazolium, pyridinium, piperidinium, or pyrrolidinium with alkyl side chains) and organic or inorganic anions. Presently, the most commonly used ionic liquids are 1-alkyl-3-methylimidazolium [Cnmim] salts. Ionic liquids are used as solvents instead of the hazardous organic solvents, and are considered to be much more environmentally friendly than those. The "environmentally friendly" affix is mainly associated with them having very low volatility, whereby any air pollution caused by them is markedly reduced. However, the compounds will be distributed in the aqueous environment, and can thus be potentially toxic to aquatic biota. The potential toxicity is increased as they are poorly degraded (by microbes). The present information suggests that ionic liquids are moderately toxic, but since their toxicity depends on the ions they consist of, and since it is poorly known what their concentrations in natural waters can be, their environmental impact is presently uncertain. At least one of their mechanisms of toxic action is that they interfere with antioxidant defense.

2.5 NANOMATERIALS

During recent years, nanomaterials have become the most studied entity within toxicology. Nanomaterials are defined as having diameter of 1–100 nm. To be considered a nanomaterial, at least 50% of all the particles in the material must have such a diameter. These materials are increasingly used in various industrial applications, presently in more than 1000 different products, ranging from foodstuffs to solar panels. The problem in characterizing their environmental effects is that their form in the environment is poorly known. Their aggregation state can change markedly depending on the conditions. This affects both their bioavailability and general toxicity. The size and sedimentation of nanomaterials depend on the movement of water in the vicinity of the material. Consequently, markedly different toxicities are obtained in static and flow-through toxicity tests and, generally, traditional toxicity testing gives a poor picture of their effects. Thus, developing methodologies for toxicity testing of nanomaterials is of major importance. Most presently used nanomaterials are metal-based, and one question, which is only partially clarified, is to what extent any measured toxicity is due to the metal ions and what is due to the metal in the nanomaterial itself. Because metal-based nanomaterials exist, in principle, not as metal ions, their uptake in organisms is different from dissolved aquatic metal ions. They can be taken up by pinocytosis in the gills or the gut (together with components of food). The uptake route may affect the observed toxicity in the organism. One of the toxic effects of nanomaterials is the inflammation they can cause on contact surfaces. Another suggested toxicity mechanism is oxidative stress. The reason for oxidative stress is presently unknown. Some carbon nanotubes cause disturbances similar to those observed in asbestos exposure. One possible reason for such a response has been suggested to be the production of ROS by phagocytes in their futile attempt to destroy the nanotubes. In addition to metal and carbon-tube nanoparticles, metal oxide nanoparticles, such as titanium dioxide in sunscreens and cosmetics and manganese oxide in industrial nanomaterial production, are quite common. The coating of nanomaterials may affect their toxicity: when titanium dioxide nanoparticles, which appear only slightly toxic, were coated with silica dioxide, the toxicity was increased. In addition to the intentional coating, the interaction of nanomaterials with biomolecules, such as sugars, lipids, and proteins, generates additional coatings, and can influence the bioavailability and toxic effects of the compounds. Thus, one must conclude that the distribution and toxic effects of nanomaterials in aquatic systems are poorly known, one problem being that new methodologies must be developed before relevant toxicity testing on these materials can be done.

2.6 RADIATION

2.6.1 Radioactivity

There are three major types of radioactive emissions: α, β, and γ radiation (Figure 2.4). Alpha radiation (which is caused by the transport of helium nuclei) has the highest energy, but its penetration is small because of the large size of the emitted particles. The thickness of skin is adequate to reduce the radiation to a negligible level. However, whenever the particles are transported into cells they cause damage, such as chromosome breakage, protein denaturation, and membrane leakage. The alpha-emitting isotopes, e.g. plutonium-238, are consequently very dangerous. Beta radiation is the movement of electrons or positrons. The energy of beta particles varies markedly: it is very small for tritium and sulfur, but quite large in the case of, for example, phosphorus and sodium. All of these isotopes are commonly used in biological laboratories. The most common protection against the radiation is the use of plexiglass shielding more than 1.5 cm thick. The energy of gamma radiation is very small, but, because of the virtually nonexistent size of particles, its penetration distance is long. Consequently, the likelihood that the radiation causes cellular damage is small, but since the effective distance is long, some cells in an organism may be damaged. As protection against γ radiation, lead tiles are commonly used. In particular, the isotopes of iodine deserve attention, since they are in common use in biological laboratories and since they

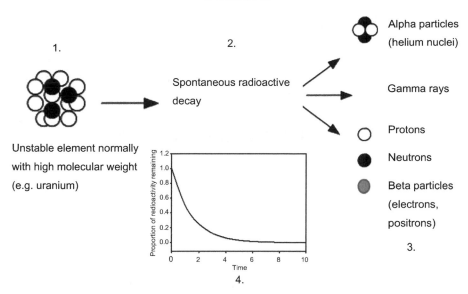

FIGURE 2.4 **A schematic picture of radioactive decay.** (1) The nuclei of several isotopes (especially of high-molecular-weight elements such as uranium) are unstable and undergo (2) spontaneous radioactive decay to elements with smaller molecular weight. The intermediates may continue to decay, but the final product (isotope of an element) is stable. (3) During the decay, essentially three types of radioactive emission (α, β, and γ) are given out. (4) The radioactive toxicity depends on, in addition to the type of emission and its energy, the half-life of the isotope. The figure gives the radioactivity remaining as a function of time. The half-life of the isotope is the time point where 50% of the radioactivity remains.

accumulate in the thyroid gland. In addition to the radiation type (particle size) and its energy, the half-life (the time taken for the radiation of a source to decrease by 50%) is an important aspect in radioactive pollution. The half-life varies from fractions of a second to thousands of years. Naturally, other things being similar, the isotope with the longer half-life is the more hazardous.

The sources of radioactivity to aquatic environments are nuclear power plants, nuclear-waste-treatment plants, uranium mine effluents, and radioactive (radium-contaminated) groundwater. Contaminated groundwater is a problem in wells in some granite bedrock areas. Nuclear-waste-treatment plants, which produce plutonium for military purposes, have caused significant marine contamination. The most notable example of such cases is the British Sellafield. In contrast, as long as nuclear power plants are functioning properly, the radioactivity escaping to the environment remains small and no effects on organisms are observed. Normal coal power plants give out more radioactivity than a well-functioning nuclear power plant. However, the radioactive material needs to be mined and transported to the place of use, where the radioactive waste must first be stored and then transported to the final storage site. All of these steps can cause aquatic pollution. Also, in the case where power plants are built in areas where earthquakes may take place, damage to power plants and consecutive radioactive pollution, as recently observed in Japan, may occur. Nuclear power plants are virtually always built in the vicinity of water bodies because of the need for adequate cooling water. Thus, any environmental pollution is likely to be aquatic.

2.6.2 Ultraviolet Radiation

Ultraviolet (UV) radiation has increased because of stratospheric ozone depletion. In aquatic systems, the radiation may cause the generation of ROS and oxidative stress in organisms. Also, UV radiation can cause transformation of chemicals. The effects are restricted to surface water because the penetration of UV radiation in water is limited.

2.7 GENETIC MODIFICATION

A modified gene functions similarly regardless of whether the genetic modification is achieved by traditional selection or by molecular methods. Since the effects depend on the gene modified, the actual effects on the function of organisms cannot be discussed in a general fashion. In principle, using gene transfer to produce genetically modified organisms is a powerful method, as properties that would take hundreds of generations to achieve or that cannot be obtained at all using traditional breeding can be obtained in a single generation. However, there are some general questions that must be addressed. The first of these concerns how the genetically modified organisms are selected. Often, the selection is done using antibiotic resistance. In this case, a gene conferring antibiotic resistance is close to the gene of interest. Consequently, only organisms having undergone successful gene transfer will grow in antibiotic-containing medium. This mechanism may increase the development of antibiotic resistance. Large-scale gene transfer has commonly been done using methods that do not allow accurate determination of the placement of the transferred gene in the genome. Consequently, the possibility remains that, even when gene transfer is successful, the transferred gene is situated within the regulatory regions of other genes. If this is the case, the function of these other genes may be disturbed. Further, it cannot be certain that the transferred gene will not jump to closely related wild organisms. This possibility is particularly important to take into account with genetically modified plants. The possibility that genetically modified organisms will escape from their rearing place and become established in an ecosystem must always be considered. In the case of transgenic fish, this possibility has been prevented by making the transgenic fish sterile. The possibility that transgenic animals might have a competitive advantage over naturally reproducing ones stems from the idea that animals prefer bigger partners as their mates. This would favor transgenic animals, as the single trait that has been most selected for is growth. However, this notion has not been conclusively tested with transgenic animals.

Relevant Literature and Cited References

Ansari, A.A., Gill, S.S., Lanza, G.R., Rast, W. (Eds.), 2010. Eutrophication: Causes, Consequences and Control. Springer, Berlin.

Beyer, W.N., Meador, J.P. (Eds.), 2011. Environmental Contaminants in Biota: Interpreting Tissue Concentrations, second ed. CRC Press, Boca Raton, FL.

Burton, D.L., Hall, T.J., Fisher, R.P., Thomas, J.F. (Eds.), 2004. Pulp and Paper Mill Effluent Environmental Fate and Effects. DEStech Publications, Lancaster, PA.

Calow, P. (Ed.), 2009. Handbook of Ecotoxicology. Wiley, Hoboken, NJ.

Clark, E.A., Sterritt, R.M., Lester, J.N., 1988. The fate of tributyl tin in the aquatic environment. Environ. Sci. Technol. 22, 600–604.

Committee on Understanding Oil Spill Dispersants: Efficacy and Effects, National Research Council, 2005. Oil Spill Dispersants: Efficacy and Effects. National Academies Press, Washington, DC.

Craig, P.J., 2003. Organometallic Compounds in the Environment. Wiley, Hoboken, NJ.

Daughton, C.G., Ternes, T.A., 1999. Pharmaceuticals and personal care products in the environment: Agents of subtle change? Environ. Health Perspect. 107 (Suppl. 6), 907–938.

Fent, K., Weston, A.A., Caminada, D., 2006. Ecotoxicology of human pharmaceuticals. Aquat. Toxicol. 76, 122–159.

Fingas, M., 2012. The Basics of Oil Spill Cleanup, third ed. CRC Press, Boca Raton, FL.

Harrad, S., 2009. Persistent Organic Pollutants. Wiley-Blackwell, Hoboken, NJ.

Lavoie, R.A., Jardine, T.D., Chumchal, M.M., Kidd, K.A., Campbell, L.M., 2013. Biomagnification of mercury in aquatic food webs: A worldwide meta-analysis. Environ. Sci. Technol. 47, http://dx.doi.org/10.1021/es403103t.

Logsdon, M.J., Hagelstein, K., Mudder, T.I., 1999. The Management of Cyanide in Gold Extraction. International Council on Metals and the Environment, Ottawa, Ontario.

Mason, R.P., 2013. Trace Metals in Aquatic Systems. Wiley-Blackwell, Hoboken, NJ.

Mwangi, J.N., Wang, N., Ingersoll, C.G., Hardesty, D.K., Brunson, E.L., Li, H., Baolin, D., 2012. Toxicity of carbon nanotubes to freshwater aquatic invertebrates. Environ. Toxicol. Chem. 31, 1823–1830.

Nordberg, G.F., Fowler, B.A., Nordberg, M., Friberg, L.T. (Eds.), 2007. Handbook on the Toxicology of Metals, third ed. Elsevier, Amsterdam, The Netherlands.

Sahu, S.C., Casciano, D.A. (Eds.), 2009. Nanotoxicity: From In Vivo and In Vitro Models to Health Risks. Wiley, Hoboken, NJ.

Sanchez-Bayo, F., van den Brink, P.J., Mann, R.M. (Eds.), 2013. Ecological Impacts of Toxic Chemicals. Bentham Science, Sharjah, United Arab Emirates.

Sumpter, J.P., 2005. Endocrine disrupters in the aquatic environment: An overview. Acta Hydrochim. Hydrobiol. 33, 9–16.

Tjeerdema, R.S. (Ed.), 2012. Aquatic life water quality criteria for selected pesticides. Rev. Environ. Contam. Toxicol. 216.

Vost, E.E., Amyot, M., O'Driscoll, N. (Eds.), 2011. Environmental Chemistry and Toxicology of Mercury. Wiley-Blackwell, Hoboken, NJ.

Webber, C.D., 2010. Eutrophication: Ecological Effects, Sources, Prevention and Reversal. Nova Science, Hauppauge, NY.

Wood, C.M., Farrell, A.P., Brauner, C.J. (Eds.), 2011. Homeostasis and Toxicology of Essential Metals. Fish Physiology, Volume 31, Part A. Academic Press, San Diego, CA.

Wood, C.M., Farrell, A.P., Brauner, C.J. (Eds.), 2011. Homeostasis and Toxicology of Non-essential Metals. Fish Physiology, vol. 31, Part B. Academic Press, San Diego, CA.

Zalups, R.K., Koropatnick, D.J. (Eds.), 2000. Molecular Biology and Toxicology of Metals. CRC Press, Boca Raton, FL.

Principles of Water Purification

Abstract

This chapter describes the principles of water purification. An important point to observe is the difference between purifying drinking water and purifying water optimal for life of organisms. In the former case, it is important that organisms, prokaryotes and protists, are effectively killed in the water treatment. In the latter instance, the purified water must allow all organisms to live. Chlorination and other treatments that are used to purify drinking water are toxic to all organisms. Water treatment first mechanically removes large objects, whereafter much of the organic material is biodegraded via digestion by anaerobic and aerobic bacteria. When wastes are biodegraded, production of biogas and heat occurs. A final step in wastewater treatment involves the removal of certain compounds, such as phosphorus by precipitation as, for example, insoluble iron phosphate, and of some metals by hyperaccumulating plants.

Keywords: wastewater-treatment plant (WWTP); sludge; primary treatment; secondary treatment; biogas; biofilm; biological oxygen demand (BOD); chemical oxygen demand (COD).

3.1 PRINCIPLES OF WASTEWATER TREATMENT

Purification of water is described schematically in Figure 3.1. In pretreatment, the incoming wastewater can be coarsely filtered to remove large and easily removable particles such as tree branches, plastic bags, etc. In pretreatment, the wastewater can also flow through channels where material such as sand and stones sinks to the bottom and can be removed, so it does not break the pumps and other mechanical constituents in further wastewater treatment.

Removal of large objects is continued in primary treatment, which typically consists of holding the wastewater in quiescent basins, primary settling tanks, where heavy objects and material

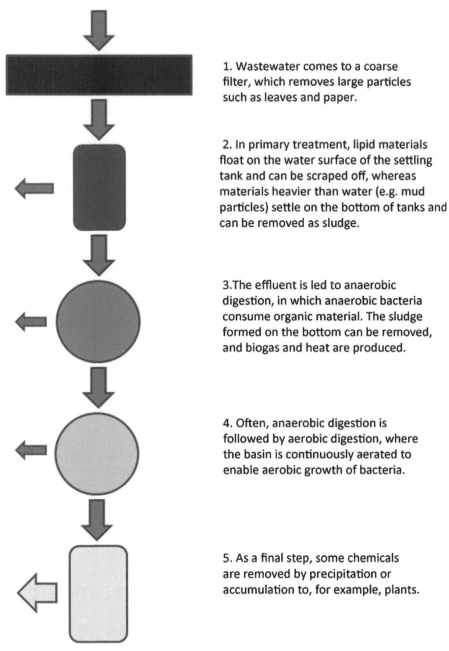

1. Wastewater comes to a coarse filter, which removes large particles such as leaves and paper.

2. In primary treatment, lipid materials float on the water surface of the settling tank and can be scraped off, whereas materials heavier than water (e.g. mud particles) settle on the bottom of tanks and can be removed as sludge.

3. The effluent is led to anaerobic digestion, in which anaerobic bacteria consume organic material. The sludge formed on the bottom can be removed, and biogas and heat are produced.

4. Often, anaerobic digestion is followed by aerobic digestion, where the basin is continuously aerated to enable aerobic growth of bacteria.

5. As a final step, some chemicals are removed by precipitation or accumulation to, for example, plants.

FIGURE 3.1 **A schematic representation of the different components in a wastewater-treatment plant.** The strength of the blue color indicates the proportion of waste in the water. The lighter the color the cleaner the effluent. Green arrows indicate that the sludge formed can be removed for the various forms of sludge treatment.

with higher density than water settle to the bottom of the tank, and light objects, oil, and grease float on the surface. Primary treatment consists of removing both of these. The bottoms of primary settling tanks commonly contain mechanical devices that move the formed sludge forward to hoppers in the base of tank, where it can be removed to sludge-treatment facilities.

The water then continues to flow to secondary treatment. The main purpose of secondary treatment is to decrease the amount of organic material, among it nutrients and detergents, from wastewater. Basically, this can be done by allowing organisms to metabolize any organic material from the wastewater. The simplest system for doing this is a series of ponds with healthy communities of organisms. The amount of organic material decreases from pond to pond. A similar system is subsurface-flow constructed wetland, where wastewater is pumped to and resides a minimum of four days in the wetland, which has all the components of a wetland community. In both of the above systems, macroscopic components of the communities are present, and any formed sediment/soil/sludge cannot be removed. Additionally, the efficiency of water cleaning depends very much on the ambient weather. In the temperate zone, very little respiration occurs in the cold temperatures of winter, decreasing the efficiency of water cleaning by these secondary-treatment methods. The most common secondary-treatment methods involve the use of bacteria, protozoa, and other protists in water-cleaning basins. Typically, the organisms are indigenous microorganisms. Two types of system are used: fixed-film (attached-growth) and suspended-growth systems. In the first type, the organic-material-consuming organisms form biofilms on solid supports. To increase the amount of biofilms, the support site area (number) is increased. Systems used include biotowers and rotating biological contactors (see Figure 3.2). The sewage moves past the biomass that is attached to the solid supports. In suspended-growth systems, such as the activated sludge system, the biomass is mixed with the sewage. The sludge that sediments on the bottom of the basin can usually be removed for further sludge treatment via bottom

(A)

(B)

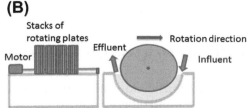

FIGURE 3.2 **Principles of biotowers and rotating biological contactors.** (A) In biotowers, the wastewater is pumped into the top of the tower, and trickles down from plate to plate. Biofilms are formed on the plates, which serve to increase the area of biofilm using up the organic carbon of wastewater as the energy source of respiration. (B) In rotating biological contactors, biofilms are formed on the plates, which are rotated by a motor. The wastewater is led to basins where the water level is at about 40% of the disc height, which is enough to ensure that the rotating disc is moist throughout the rotation, whereby the organic material in the wastewater can be used up by the biofilms.

valves. Secondary treatment often consists of the use of both anaerobic and aerobic bacteria. To guarantee the growth of the latter, the basin is aerated (which is not done for basins where anaerobic growth is wanted). Typically, the first phase of secondary treatment consists of anaerobic bacteria metabolizing the wastewater. Thereafter, the water is bubbled with air in new tanks to enable the growth of aerobic bacteria. The suspended-growth systems can better accommodate drastic changes in water-treatment requirements than fixed-film systems, as more biomass can be added to the system as needed.

An important aspect of water cleaning is the removal of metals. To do this, metal uptake by plants has been intensively studied. If a plant takes up a metal effectively, it is first allowed to accumulate the metal and then harvested from the pond, leaving metal-depleted water behind. Many metals can also be precipitated with chemical addition. Both metal-removal methods belong to tertiary treatment, which refers to any further steps of water purification (after secondary treatment) before the water is allowed to flow to discharge sites in fragile ecosystems. As described above for metal depletion, the tertiary treatments may consist of addition of chemicals to precipitate contaminants, or they may involve disinfection of water by ultrafiltration through sand or activated carbon or allowing it to stay in man-made lagoons. Presently, phosphorus, for example, is largely removed from wastewater by precipitating ferric phosphate: $FeCl_3$ is added to the sewage, whereby $FePO_4$ is formed and precipitated as the salt is sparingly soluble.

The efficiency of modern wastewater-cleaning units is relatively high: the European water directive requires that more than 80% of the phosphorus is removed. The efficiency of phosphorus removal can be much higher—some sewage treatment plants can remove more than 95% of the phosphorus contained in wastewater. The efficiency of nitrogen removal is smaller: 50–80% is typically removed. Whereas phosphorus remaining after secondary treatment is largely removed by precipitation, nitrogen removal depends solely on the efficiency of biological treatment. Initially, bacteria convert ammonium nitrogen (NH_3) to nitrite, which is further converted to nitrate and nitrogen gas by other bacteria. The gas is then given up to the atmosphere. Ammonia is decreased much more than total nitrogen, with removal efficiency ranging from 80 to more than 95%. The biological oxygen demand (BOD), which is one of the major parameters given to indicate the quality of (waste)water, decreases more than 90% in efficient cleaning. The chemical oxygen demand (COD), also a major measured water-quality parameter, typically decreases by more than 95%. The COD and BOD are parameters that are commonly used as determinants of water-treatment efficiency. Total organic carbon (TOC) is decreased by more than 90% in efficient wastewater-treating facilities, with more than 85% decrease in dissolved organic carbon (DOC) concentration. Major problems for water treatment are pharmaceuticals. There are two reasons for this. First, pharmaceuticals are manufactured to resist unwanted degradation, whereby they can pass through the water treatment and be released to the environment unmodified. The acronym EPPP (environmental persistent pharmaceutical pollutant) has consequently been coined. Second, pharmaceuticals are often bactericides, and may thus kill the bacteria that perform most of water purification in secondary treatment, decreasing the efficiency of water purification.

3.2 DISINFECTION STEPS FOR GENERATING HOUSEHOLD WATER

In addition to the above, water purification for drinking/household water contains additional steps. Good household water should be free from bacteria and other harmful organisms.

However, the disinfection methods utilized are also toxic to nontarget organisms. Thus, safe drinking water may kill, for example, aquarium fish. Disinfection methods for making household water are:

1. Chlorination. This is the most common form of (waste)water disinfection, as it is cheap and effective. Adding ammonia and chlorine simultaneously generates chloramines. Chloramines can also be added as a compound to drinking water pipes. Chloramines are not used in the treatment of wastewater because they are highly persistent, which makes them good for disinfecting drinking water pipes, as the effect remains throughout the piping system. Chlorination can cause the formation of carcinogenic or otherwise harmful halogenated compounds from the organic material that the water contains. Because residual chlorine is toxic to aquatic organisms, the water should be chemically dechlorinated before it is allowed to enter the environment.
2. Ultraviolet-light treatment. The water to be disinfected is led past ultraviolet (UV) light lamps. As no chemicals are used in disinfection, the treated water has no adverse effects on organisms. The major disadvantage of UV disinfection is the need for frequent lamp maintenance and replacement. Also, since the disinfection is local, any harmful microorganisms in the water pipes after the disinfection site will not be affected.
3. Ozonation (O_3 treatment). Ozone is generated from normal molecular oxygen at high voltages. Ozone oxidizes most organic materials, thereby destroying most pathogens. Ozone is safer than chlorine, because it produces fewer harmful by-products, and because ozone is generated on-site as needed, whereas chlorine must be stored as toxic gas. A disadvantage of ozone disinfection is the high cost of the ozone-generation equipment.

One additional problem with any wastewater-treatment unit is the space that the facility needs. Also, in temperate areas, the facilities must be warmed to guarantee efficient water cleaning in winter at cold temperatures.

3.3 SLUDGE TREATMENT

Although the treatment of the sludge formed in the primary and especially in the secondary treatment is strictly speaking not water purification, it belongs to a complete treatment of wastewater purification. The sludge treatment and disposal can proceed in the following way. The sludge contains organic matter, metals, and a number of disease-causing microorganisms. Before the sludge can be utilized, these must be removed as effectively as possible. The most common treatment options include anaerobic digestion, aerobic digestion, and composting. Incineration is also occasionally used. Small facilities often use composting, midsized ones aerobic digestion, and large ones anaerobic digestion. Occasionally, the liquid in the sludge is decreased by centrifugation (which includes both centrifugal and rotary-drum sludge thickeners) or by the use of belt filter presses. In anaerobic digestion, the sludge is fermented in tanks with high (55 °C—thermophilic) or moderate (36 °C—mesophilic) temperature in the absence of oxygen. The biogas (methane) produced in anaerobic digestion can be used in energy production (heating, electricity), increasing the environmental friendliness of the treatment. For example, many vehicles running with natural gas are nowadays actually using biogas liberated in wastewater treatment. Aerobic digestion uses aerobic bacteria. These rapidly consume organic matter and convert it into carbon dioxide. Aerobic digestion is relatively expensive as energy is required for the blowers, pumps, and motors needed to add oxygen to the process. However, much of the energy needed can be produced by the treatment plant itself via biogas and heat production. The most cost-efficient method to achieve aerobic digestion is probably the use of fine bubble diffusers, where gas enters the sludge through small holes. The problem is that the small aeration holes easily get plugged by sediment. Composting is also an aerobic process. In this method, the wastewater sludge is mixed with sources of carbon such as sawdust, straw, leaves, or plant remains. Bacteria digest both the sludge and the added carbon source, and produce a large amount of heat. Incineration of sludge is not often used because of the air emissions produced and the supplemental fuel needed to burn the sludge and to vaporize the residual water it contains. The treated dry sludge can be used as fertilizer. However, if the sludge contains toxic compounds (including metals) it cannot be used in agriculture. Such sludge is either used as landfill or incinerated in hazardous-waste facilities.

Relevant Literature and Cited References

Baird, R., Smith, R.K., 2002. Third Century of Biochemical Oxygen Demand. Water Environment Federation, Alexandria, VA.

Bouki, C., Venieri, D., Diamadopoulos, E., 2013. Detection and fate of antibiotic resistant bacteria in wastewater treatment plants: A review. Ecotoxicol. Environ. Saf. 91, 1–9.

Choi, S., Dombrowski, E.-M. (Eds.), 2007. Fundamentals of Biological Wastewater Treatment. Wiley-WCH, Weinheim, Germany.

Cohen, J.M., Kugelman, I.J., 1972. Wastewater treatment: Physical and chemical methods. J. Water Pollut. Control Fed. 44, 915–923.

Copp, J.B., Spanjers, H., Vanrolleghem, P.A., 2002. Respirometry in Control of the Activated Sludge Process: Benchmarking Control Strategies. IWA, London.

Eduok, S., Martin, B., Villa, R., Nocker, A., Jefferson, B., Coulon, F., 2013. Evaluation of engineered nanoparticle toxic effect on wastewater microorganisms: Current status and challenges. Ecotoxicol. Environ. Saf. 95, 1–9.

Forster, C., 2003. Wastewater Treatment and Technology. Thomas Telford, London.

Lettinga, G., 1995. Anaerobic digestion and wastewater treating systems. Antonie van Leeuwenhoek 67, 3–28.

Libhaber, M., Orozco, A., 2012. Sustainable Appropriate Technologies for Treatment of Municipal Wastewater. IWA, London.

Kadlec, R.H., 2008. Treatment Wetlands, Second ed. CRC Press, Boca Raton, FL.

Kagle, J., Porter, A.W., Murdoch, R.W., Rivera-Cancel, G., Hay, A.G., 2009. Biodegradation of pharmaceutical and personal care products. Adv. Appl. Microbiol. 67, 65–108.

Kelessidis, A., Stasinakis, A.S., 2012. Comparative study of the methods used for treatment and final disposal of sewage sludge in European countries. Waste Manag. 32, 1186–1195.

Michael, I., Rizzo, L., McArdell, C.S., Manaia, C.M., Merlin, C., Schwartz, T., Dagot, C., Fatta-Kassinos, D., 2013. Urban wastewater treatment plants as hotspots for the release of antibiotics in the environment: A review. Water Res. 47, 957–995.

Sahm, H., 1984. Anaerobic wastewater treatment. Adv. Biochem. Eng. Biotechnol. 29, 83–115.

Tahar, A., Choubert, J.M., Coquery, M., 2013. Xenobiotics removal by adsorption in the context of tertiary treatment: A mini review. Environ. Sci. Pollut. Res. Int. 20, 5085–5095.

Tunay, O., 2003. Developments in the application of chemical technologies to wastewater treatment. Water Sci. Technol. 48, 43–52.

Vymazal, J., 2011. Constructed wetlands for wastewater treatment: Five decades of experience. Environ. Sci. Technol. 45, 61–69.

Zhu, G., Peng, Y., Li, B., Guo, J., Yang, Q., Wang, S., 2008. Biological removal of nitrogen from wastewater. Rev. Environ. Contam. Toxicol. 192, 159–195.

Sources and Transport of Chemicals in Aquatic Systems

Abstract

This chapter first outlines the major sources of aquatic contamination. Contamination can be accidental or deliberate. Point sources include industrial and household effluents, which are often cleaned in wastewater-treatment plants (WWTP), and enter recipient waters in effluent pipes. Particularly in the past, but even nowadays, effluents can be discharged into aquatic bodies with virtually no cleaning. This minimal cleaning is a deliberate choice, usually taken to reduce the expenses incurred in effluent treatment. It is pointed out that as movements of water do not follow national boundaries, strict environmental standards in one nation are not effective to keep the waters of that nation clean, if other areas that are situated along the waterways have slack environmental standards. Accidental releases of contaminants are caused especially by malfunctioning wastewater-treatment plants, shipwrecks, and accidental spills in oil production. Such accidental spills are on the increase as more and more oil drilling takes place on oceanic coastal floors. In contamination by diffuse sources, agriculture plays the major role. To decrease agricultural contamination, safety zones near waterways and temporal planning of, for example, nutrient application, is important. Also, avoidance of excess fertilization, and technological advances in contaminant (nutrients and pesticides) application help in decreasing contamination. Once in the waterways, contaminants can travel long distances in rivers and on ocean currents. Airborne contaminants can be carried in air currents, and enter waterways as a result of rain.

Keywords: point source; diffuse source; leaching; runoff; fertilizers; water current; air current; accidental spill; deliberate spill; safety zone; waste dumping.

An Introduction to Aquatic Toxicology
http://dx.doi.org/10.1016/B978-0-12-411574-3.00004-9

4.1 THE MAJOR SOURCES OF POLLUTANTS

Contaminants reach aquatic bodies either deliberately or by accident. An overview of the sources/reasons of aquatic contamination is given in Figure 4.1. The first major source of aquatic contamination is household and industrial wastewater. Although this wastewater could undergo reasonable effective treatment (see Chapter 3), there are marked differences between countries in the requirements for water cleaning before disposal to the receiving environment, be it lake, river, or ocean. The pronounced differences in water-cleaning requirements between countries is unfortunate, as contaminants in rivers and oceans do not distribute according to national borders, so it is possible that the shores of a nation with strict requirements are polluted from a source in a nation not requiring effective wastewater treatment (see Figure 4.1). However, in principle, point sources, whether communities or industry, can be covered by wastewater-treatment facilities. Thereafter, the major problem is to account for contaminants that are not broken down or detoxified in wastewater treatment.

A major group of such contaminants is pharmaceuticals, which are often designed to be resistant to unwanted breakdown. Another problem is diffuse sources, which are not associated with central water-purification systems. This pertains especially to small rural villages and individual houses.

The second significant source of contaminants is leaching from land. This is the way that agricultural contaminants, both fertilizers and pesticides, reach aquatic systems. Leaching of both depends on the precipitation during the time of application: the more it rains, the more contaminants will reach water bodies. In areas covered with snow, although it used to

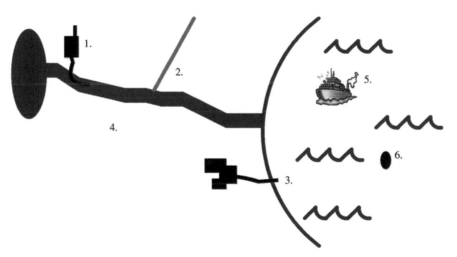

FIGURE 4.1 **Major sources of aquatic pollution.** The point sources include both industry (1) and households (towns; 3). The effluent may be dispersed to rivers, lakes, or the sea. Even if a country (in the figure, right of the red line and above the river; 2) has strict environmental standards, it will be contaminated by toxicants carried in the river or currents in the ocean. This shows that aquatic pollution is an international question and cannot be resolved within nation states. Diffuse sources include agricultural fertilizer and pesticide use (4). Any shipwrecks or intentional disposal of chemicals/garbage from ships (5) causes aquatic pollution, as does release of toxicants from sediments or leakage from earlier dumped chemical containers (6).

be customary to spread fertilizing manure on the snow covering the fields, most of the fertilizing substances ended up in water bodies as they were associated with meltwater. Leaching depends on the distance between the place of application and the water body. To reduce fertilizer runoff to water bodies (rivers, lakes, and sea), so-called "safety zones" between cultivated land and water bodies are used. In addition, the mode of application affects runoff: if the fertilizer and pesticide are directly applied in the vicinity of growing seed, much less runoff occurs than when, for example, aerial spraying of pesticides is undertaken. Notably, aerial spraying may cause direct contamination of small waterways with the pesticide, and consequent exposure of aquatic nontarget organisms to the compound. Excess fertilizer use is also not beneficial, as the fertilizer that is not used up by the cultivated plants ends up in the aquatic environment after leaching. Besides agricultural fertilizer and pesticide runoff, a much-debated source of aquatic, especially groundwater, contamination is salting of roads in areas where freezing occurs. The increased salinity of water has been observed to cause effects on tadpoles and groundwater organisms.

Aquaculture generates a significant source of fertilizing compounds, as fish feeds are not utilized to 100% and as the feces of fish contain fertilizing compounds. The fertilization caused by these fish is concentrated in a much smaller area than in the case of normal fish populations. Aquaculture also causes the appearance of antibiotics and antiparasitic drugs in organisms in the vicinity of aquaculture facilities. Pesticide use in water bodies may also be intentional as, for example, herbicides are used to keep aquatic plants in lakes in check and insecticides are used to reduce populations of possible fish parasites close to aquaculture.

Intentional dumping of hazardous chemicals to lake and sea bottoms has been quite common: the chemicals and their dumping sites are poorly known, but may include, for example, nerve gases and other chemicals for chemical warfare from the First and Second World Wars. To limit the contamination of surface waters, areas with great depth have often been chosen as dumping sites. However, because of the high pressure in deep water, the containers of toxicants are likely to start leaking earlier than in shallower waters. In such cases, although surface water may be saved from contamination (if circulation of water between depths is limited, as often is the case, especially in oceans), toxic effects may take place in deepwater organisms. However, although chemical containers in shallow water remain intact longer than in deep water, ultimately they start to leak, and poor knowledge of the nature of the chemicals and their dumping sites prevents any protective action. Some small-scale oil pollution is also intentional, and caused by cleaning of tanks for further use. However, most of the oil and chemical spillages are accidental, caused by shipwrecks or accidental spills during oil exploration.

Another major source for polluting compounds is aerial deposition. The discharges from vehicles, and all combustion smoke contain toxic compounds. For example, much of the dioxin load in the environment is due to incomplete combustion. Smoke is the reason for acid rain, which causes the acidification of waterways. The smoke gases contain sulfur and nitrogen oxides, with the consequence that sulfurous, sulfuric, and nitric acids are deposited in waterways as a result of rain. In Europe and North America, the amount of sulfur in fuel has decreased, and cleaning of smoke is required, with the result that acidification of freshwater as an environmental problem has all but disappeared (see section 1.1 for a description of acid rain production).

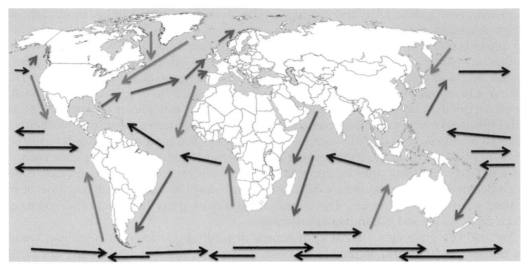

FIGURE 4.2 **An overview of the major oceanic currents.** (Coastal) cold currents are indicated in blue, and warm currents in red. As a result of oceanic currents, chemicals can be diluted and transported far away from the original place of disposal. This is most clearly seen in the North Pacific, North Atlantic, and Indian Ocean garbage patches, where garbage from thousands of miles away is found in high amounts.

Finally, bottom sediments are a major source of aquatic contaminants. In some places, contaminant release from them can be greater than current discharges. Because of this, even if effluent discharge was completely stopped, a water area where earlier wastewater has been released may remain polluted for tens of years before clear improvement of environmental conditions takes place. Typically, the polluted bottom sediments are anoxic with high concentrations of the toxic hydrogen sulfide. Furthermore, they commonly contain large deposits of phosphorus and nitrogen (in various compounds), which contribute to continued eutrophication when liberated from the sediment into the water.

4.2 TRANSPORT OF POLLUTANTS IN THE ENVIRONMENT

Contaminants can be transported long distances from the site of production to the site of measurement. It is, for example, possible that a compound produced by a factory in central Europe finds its way to polar bear tissues on the Norwegian island of Spitsbergen. The long-distance transport of contaminants depends on air and water currents. The entrance of compounds from air currents to water naturally requires that the compound associates with rain. The principles of xenobiotic transport in major ocean and air currents are given in Figures 4.2 and 4.3. Because of ocean currents, waste items such as plastic bags have accumulated on the open sea. Three major areas of garbage accumulation occur: the North Pacific garbage patch, which is currently the biggest one; the North Atlantic garbage patch; and the Indian Ocean garbage patch. Aerial deposition causes a large proportion of aquatic contamination. As an example, up to 40% of the nitrogen compounds causing eutrophication of waters is from aerial

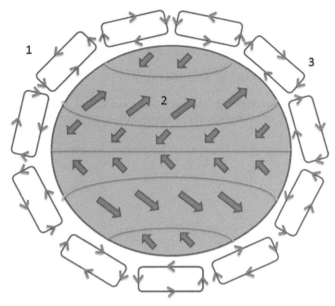

FIGURE 4.3 **An overview of typical air currents on the globe.** (1) The air flows so that the direction of the flow near the ground is opposite to that high up in the air. Where the air rises, it cools and water may condense leading to rain. (2) Typical wind directions: Polar easterlies, temperate westerlies, and northeastern and southeastern trade winds. (3) Persistent organic pollutants volatilize in warm air, rise up in the air column, and are transported along the air currents to colder areas where they condense and are precipitated (in rain). The pollutants precipitated on land may leach into the ocean and other aquatic bodies.

deposition. A significant component of long-range contaminant transport is that migrating animals may accumulate a contaminant at one end of their migration route, and it is then obtained by predators preying on the animal at the other end. Another point, related to the long-distance transport of halogenated hydrocarbons, is that although their production has decreased (see section 2.4.6), they may be re-emitted from contaminated temperate ecosystems during particularly warm spells, carried volatilized as long as the air remains warm, and become redeposited in the cold temperatures of high latitudes. This type of effect has caused peak concentrations of persistent organic pollutants (POPs) to occur in the Arctic up to tens of years after their use has been banned (in Europe and North America).

Relevant Literature and Cited References

Bignert, A., Olsson, M., Persson, W., Jensen, S., Zakrisson, S., Litzen, K., Eriksson, U., Häggberg, L., Alsberg, T., 1998. Temporal trends of organochlorines in Northern Europe, 1967–1995: Relation to global fractionation, leakage from sediments and international measures. Environ. Pollut. 99, 177–198.

Committee on the Significance of International Transport of Air Pollutants, Board on Atmospheric Sciences and Climate, Division of Earth and Life Studies, 2009. Global Sources of Local Pollution: An Assessment of Long-Range Transport of Key Air Pollutants to and from the United States. National Academies Press, Washington, DC.

Jorgensen, B.B., Richardson, K. (Eds.), 2013. Eutrophication in Coastal Marine Ecosystems. American Geophysical Union, Washington, DC.

Lehtoranta, J., Ekholm, P., Pitkänen, H., 2009. Coastal eutrophication thresholds: A matter of sediment microbial processes. Ambio 38, 303–308.

Liang, Q., Jaegle, L., Jaffe, D.A., Weiss-Penzias, P., Heckmann, A., Snow, J.A., 2004. Long-range transport of Asian pollution to the northeast Pacific: Seasonal variations and transport pathways of carbon monoxide. J. Geophys. Res. 109, D23S07.

Schnoor, J.L., 1996. Environmental Modeling: Fate and Transport of Pollutants in Water, Air, and Soil. Wiley, Hoboken, NJ.

Swackhamer, D.L., Paerl, H.W., Eisenreich, S.J., Hurley, J., Hornbuckle, K.C., McLachlan, M., Mount, D., Muir, D., Schindler, D., 2004. Impacts of atmospheric pollutants on aquatic ecosystems. Issues Ecol. 12, 1–24.

The Most Important Experimental Designs and Organisms in Aquatic Toxicology

Abstract

This chapter introduces the most important experimental designs and organisms used in aquatic toxicology. With regard to the organisms used, the benefits and problems of using only a handful of model species are discussed. As a major benefit, one can state that using the same organism in various studies enables easy comparison of work carried out by different laboratories. A major drawback is that often the results obtained may not be ecologically relevant. The availability of model organisms relevant especially for tropical and polar ecosystems is limited. With regard to experimental designs, the simplest ones are studies with a single pure toxicant and a single species in a laboratory setting. These types of studies are useful and required for understanding the modes of action of chemicals, but their environmental relevance is often small as they fail, for example, to include chemical, abiotic, and biotic interactions in the experimental design. To increase the environmental relevance of experimental settings, while still enabling replication, micro- and mesocosms are used. The pros and cons of using different "seeds" in the initiation of micro- and mesocosm communities are evaluated. Probably the major drawback of micro- and mesocosm experimentation is that one cannot include large and long-living organisms in the system. This can only be done experimentally with ecosystem manipulation. The largest system of experimental lakes is the Experimental Lake Area in Canada, which is currently under threat of being dismantled. In natural water bodies, biomonitoring is carried out. The monitoring efforts can use either native organisms or organisms "caged" in the environment for a specified period. What features of contamination either can clarify is discussed. It is pointed out that biomonitoring studies should always include controls from a clean environment with similar water properties apart from the contamination, to enable assigning a role for contamination in the measured responses. Also, in any aquatic biology study, one needs to remember that experimentation in any given area only gives values relevant to the water in that area.

Keywords: model organism; zebrafish; daphnia; microcosm; mesocosm; biomonitoring; caging; ecosystem manipulation; environmental risk assessment (ERA); biofilm.

5.1 MODEL ORGANISMS USED

A list of model organisms commonly used in aquatic toxicology is given in Table 5.1. The choice of organism used depends very much on whether one is performing toxicity testing, trying to evaluate generally how toxicants affect organismic function, or studying the likely disturbances of natural ecosystems or their components. In the first two cases, suitable model organisms can be used, whereas in the last case one needs to choose organisms relevant to the studied ecosystem.

One of the goals of researchers has been to combine these two angles and find model organisms relevant for the disturbances in the studied ecosystem. An important question determining the relevance of a study organism is whether the ecosystem is freshwater or marine, as there are virtually no model organisms that are found in both. Similarly, the climatic zone affects the relevance of a study organism for a specific ecosystem. As a large proportion of aquatic toxicological studies have been performed in the temperate zone, it is natural that most of the model organisms used are temperate species such as *Daphnia*; the occurrence of tropical or, in particular, Arctic species among model organisms is rare. A notable exception is that mainly tropical or subtropical small aquarium fish such as zebrafish, which is a tropical freshwater cyprinid, are commonly used. The zebrafish is not very well suited for many environmental studies in temperate environments, especially as it is very tolerant of hypoxia, which certainly affects any oxygen-dependent responses.

However, information obtained using model organisms often indicates the first directions where molecular, cellular, and organismic studies generally should be directed. Model organisms in the animal kingdom are mainly used for their biomedical study applications: any use in environmental sciences is subsidiary. A notable exception to this tendency is the sequencing of the *Daphnia pulex* genome. This sequencing effort indicated a large number of genes (one-third of all genes) that do not occur in the sequenced vertebrates. In particular, the transcription of these genes responds to environmental changes. This finding, the first in an aquatic short-generation invertebrate, suggests that in short-generation organisms environmental responses may be regulated via transcriptional induction of genes.

5.2 MICRO- AND MESOCOSMS

In the simplest case, studies of aquatic toxicology try to assess the mechanisms that toxicants may have at the organismic or cellular level. This is most conveniently done in laboratory exposures, which normally use single organisms and single pure toxicants. Furthermore, the duration of exposure is usually limited. Although this type of experimentation continues to give valuable information, it fails to impart knowledge about any interactions between

TABLE 5.1   Model Organisms Commonly Used in Aquatic Toxicology, With Some Internet Sources for Further Information

Organism	Notes
AMPHIBIA	
Xenopus sp. (*laevis* and *tropicalis*)	In particular, metamorphosis and endocrine disruption are studied. *Xenopus* genomes have been sequenced, and a wide variety of resources can be accessed at http://www.xenbase.org
FISH (MAJOR FEATURES OF ALL SPECIES CAN BE FOUND AT HTTP://WWW.FISHBASE.ORG)	
Zebrafish (*Danio rerio*)	Because of its pronounced use in biomedicine, zebrafish resources are immense, and can be accessed at http://www.zfin.org. The genome of zebrafish has been sequenced, and morpholinos (fish with manipulation of genes affecting the phenotype) are available, as are large numbers of commercial antibodies. The species is a small tropical cyprinid initially originating from north Indian rivers, but is a very common aquarium fish. It is easy to maintain and tolerates many environmental variations. In addition, it can be made to breed throughout the year
Rainbow trout (*Oncorhynchus mykiss*)	The most commonly used big fish in aquatic toxicology. Whenever blood samples are needed, rainbow trout is the species of choice. Apart from rainbow trout, the big fish commonly used in aquatic toxicology are carp (*Cyprinus carpio*) and goldfish (*Carassius auratus*). The latter two species are easy to maintain as they tolerate poor environmental conditions. Two major drawbacks affect the suitability of big fish for toxicological studies. First, they require hatchery-level conditions with big tanks and, in the case of salmonids, clean flow-through water. Second, their generation time is long. For this reason, life-cycle tests using them are time-consuming (several years may be needed), and require extensive space. Rainbow trout is the most commonly used species, because it is commonly cultivated throughout the world. The hatchery-reared fish are freshwater ones, but the species has anadromous populations (steelhead salmon) with marine adult animals
Three-spined stickleback (*Gasterosteus aculeatus*)	The three-spined stickleback is another species with a fully sequenced genome. The species is very commonly used in behavioral and evolutionary studies. It is also a species with a male-specific reproductive protein, spiggin, a glue protein that is used in nest building by males. Thus, it is well suited for studies of reproductive disturbance. This small temperate fish has both freshwater and marine populations
Fathead minnow (*Pimephales promelas*)	As most model fish, the species is in wide use, because it is small and easy to maintain with a short generation time. These properties have made it a favorite when reproductive toxicity and life-cycle toxicity tests are carried out. The species is a subtropical cyprinid
Medaka (*Oryzias latipes*)	The genomic sequence of this small species is available, and the animal is commonly used in life-cycle tests. All medaka species originate from southeast Asia, and are small (less than 5 cm) with a short generation time. Compared to zebrafish, it tolerates a wider temperature range, has a shorter generation time, and a smaller genome. Medakas also tolerate large variations in salinity, as they are originally brackish water species. Their use as a model originates from the fact that they are common aquarium fish
Guppy (*Poecilia reticulata*)	This small aquarium fish has been used in, for example, studies of bioaccumulation
Mosquito fish (*Gambusia* sp.)	The group *Gambusia* contains close to 100 small species. The actual mosquito fish (*Gambusia affinis*) becomes maximally 4 cm long. Mosquito fish have been introduced to a wide range of habitats for biological control of mosquitos. However, a disturbed ecosystem balance after introduction has frequently been observed. The species is viviparous

(Continued)

TABLE 5.1   Model Organisms Commonly Used in Aquatic Toxicology, With Some Internet Sources for Further Information—cont'd

Organism	Notes
Killifish (*Fundulus heteroclitus*)	The genetics of this small estuarine fish are particularly well characterized. It inhabits the east coast of North America from Florida to Nova Scotia. The genetics of toxicological responses can be studied with killifish, as populations with different tolerances to contamination have been described
Tilapia (*Oreochromis* sp.)	Tilapias are another group of freshwater species that are easily maintained. They are the most commonly used relatively large subtropical model fish
INSECTS	
Mosquito (e.g. *Anopheles gambiae*, *Aedes aegypti*)	On the whole, insects are not aquatic, but the larvae of some groups such as mosquitos and chironomids reside in water. Because several mosquito species are associated with human disease, the genome of *Anopheles gambiae* had been sequenced by 2002. Studies on mosquitos have included, for example, development of tolerance to insecticides
Chironomus riparius	This species and other chironomids have been studied, for example, to evaluate bioaccumulation and toxicity of metals, general bioaccumulation, and transfer of toxicants in an ecosystem
MOLLUSCS	
Mytilus sp. (*galloprovincialis*, *edulis*)	The blue mussel and its relatives have formed the basis of the marine biomonitoring program "Mussel Watch" since the 1980s. The purpose of the program is to evaluate the state of the marine environment throughout the temperate oceans. The species is also used, for example, in evaluating genetic adaptations to contamination, in reproductive toxicology, and in metal accumulation and toxicology. Of the marine invertebrates, the blue mussel is possibly the toxicologically best studied
Oyster (*Crassostrea gigas*)	Marine/brackish water subtidal oysters are cultivated, often close to marinas. Consequently, the effects of organic-tin-chloride-containing antifouling paints (TBT—tributyl tin) on reproduction were observed first in oysters
Ruditapes philippinarum	This bivalve species originates from coasts of Siberia and China, but has been introduced into North America and Europe, where it has turned out to be a highly invasive species, also tolerating low salinity. Since the species is nowadays found in all temperate coasts, it is commonly used in environmental monitoring
Great pond snail (*Lymnaea stagnalis*)	This snail is probably the most commonly used mollusc species in freshwater toxicology. It is a particularly useful model when studying the neurobiological effects of toxicants. The species is hermaphrodite
CRUSTACEANS	
Water flea (*Daphnia* sp.)	This temperate freshwater species group may be the most significant group in aquatic environmental studies. This is indicated by the fact that the genome of *Daphnia pulex* was sequenced because of the importance of the animal in environmental research. The sequencing effort showed that the genome contains many unique genes, which especially respond to environmental changes
Tigriopus sp. (*japonicus* and *californicus*)	In comparison to the predominant role of the freshwater crustaceans in aquatic toxicology, marine species have been little studied, and they have no commonly used test organisms. *Tigriopus* sp. have recently been advocated to become such species

TABLE 5.1   Model Organisms Commonly Used in Aquatic Toxicology, With Some Internet Sources for Further Information—cont'd

Organism	Notes
Hyalella azteca	The species is a small (< 1 cm) North American freshwater amphipod. It is the most abundant amphipod of North American lakes, and can tolerate large environmental changes, except for being quite sensitive to acidification, with a tolerance limit at pH 5
ANNELIDS	
Nereis sp. (particularly *virens*)	Annelids live buried in bottom sediments. For this reason, they are used particularly in sediment toxicology. Of marine species, those belonging to the genus *Nereis* are the primary study organisms in sediment toxicology. The genus has approximately 250 species
Lumbriculus variegatus	Of the freshwater species used in sediment toxicology, the oligochaete *Lumbriculus variegatus* is the most common. Its natural habitats are ponds and marshes, where it feeds on available microorganisms
SEA URCHINS	
Paracentrotus lividus	Sea urchins have been favorite study objects in developmental biology for decades. *Paracentrotus lividus* is a Mediterranean species that is quite tolerant of high salinity, and high concentrations of various contaminants. It inhabits especially rocky bottoms
Lytechinus variegatus	This is a tropical species found from North Carolina to Brazil. In some places it can attain high densities. Notably, although sea urchins are important study objects both in developmental biology and toxicology, the species identity is not as strictly defined as for other model organisms
PROTOZOA	
Tetrahymena sp. (*pyriformis* and *thermophila*)	The genus *Tetrahymena* is a group of ciliates that has been an important model organism in studies achieving several significant advances in experimental biology. For example, studies on how the cell cycle proceeds, telomere function and its regulation, and histone acetylation and its role in gene expression, and the discovery of some cytoskeletal elements, have all been clarified with the aid of these protozoans
GREEN PLANTS	
Duckweed (*Lemna* sp.)	This freshwater floating green plant occurs everywhere in the world. Although it is not originally native to Australasia or South America, it has been introduced there, and has invaded ponds and streams with a slow current. It is very rapidly growing. These properties have made duckweeds common models for aquatic toxicology studies on green plants
MACROSCOPIC ALGAE	
Ceramium tenuicorne	A major algal resource is http://www.algaebase.org, which describes the general properties of both macroscopic and microalgae. Another useful site is http://www.seaweed.ie. *Ceramium* is a genus of red algae. It is both species-rich (several hundred species) and widespread. The species, normally with a maximum size of 30 cm, live from intertidal to deepwater locations. The maximal depth depends naturally on light penetration, as the genus is photosynthetic. One of the cosmopolitan species in the genus, much used in toxicological studies, is *Ceramium tenuicorne*, which has been described from southeast Asia to the Baltic Sea
Ulva pertusa	The genus *Ulva* contains more than 100 species of green algae. These are marine or brackish water species that favor eutrophic environments. *Ulva pertusa* is a cosmopolitan species of the group

(Continued)

TABLE 5.1   Model Organisms Commonly Used in Aquatic Toxicology, With Some Internet Sources for Further Information—cont'd

Organism	Notes
MICROALGAE	
Skeletonema costatum	This diatom has distribution throughout the world, from Finland to New Zealand, indicating that it is tolerant to large variations in both salinity and temperature. It is a typical indicator of eutrophication. Although it is nontoxic, high abundance can cause oxygen lack during night-time
Phaeodactylum tricornutum	The species is another diatom with wide distribution encompassing a wide salinity range. Its genome has been sequenced, and it has been used in laboratory-based physiological studies for more than 50 years. Detailed information on this and other related species is given at http://genome.jgi-psf.org
Pseudokirchneriella subcapitata	A freshwater species commonly used as a bioindicator and in toxicology testing. It is available commercially from ATCC with recommended applications of, for example, algal growth-inhibition test and tests of algal growth potential
Desmodesmus subspicatus	The species is a freshwater green alga. It has cosmopolitan distribution and reaches high, bloom-like, densities in eutrophied lakes. It can exist both as colonies and as single cells. The switch between the forms can be triggered by changes in nitrogen and phosphorus levels
Chlamydomonas reinhardii	This unicellular green alga is one of the most important model algae. Both its nuclear genome and chloroplast genome have been sequenced, and information on the large amount of resources available can be accessed at http://www.chlamy.org
Chlorella vulgaris	The *Chlorella* genus comprises unicellular species of green algae. Melvin Calvin was awarded a Nobel prize for work on carbon dioxide assimilation in plants; the work was done using *Chlorella*
BLUE-GREEN ALGAE, OR CYANOBACTERIA	
e.g. *Synechococcus* sp.	Cyanobacteria are studied in aquatic toxicology largely because they include species that are toxic. Cyanobacterial toxins can be both hepatotoxins and neurotoxins. Much of the research has investigated the mechanisms of toxicity. Cyanobacteria can generate massive, often toxic, blooms both in lakes and brackish water environments. As cyanobacteria are normally autotrophic (photosynthetic), they have been used to elucidate the mechanism of photosynthesis. To facilitate studies, the genome of *Synechococcus* (*elongates*) has been sequenced. For a more detailed account, http://microbewiki.kenyon.edu/index.php/Synechococcus can be accessed
AQUATIC BACTERIA	
e.g. *Vibrio fischeri*	This marine bacterium is bioluminescent, and is commonly found in symbiosis with squid. The luminescence is used in toxicity testing: inhibition of bioluminescence by contaminants is commonly tested. The genomic sequence of this Gram-negative bacterium has been studied. Further information about available resources can be accessed at http://microbewiki.kenyon.edu/index.php/Vibrio_fischeri

Several of the organisms presented are used in the toxicity test protocols of the Organisation for Economic Co-operation and Development (OECD) and the International Organization for Standardization (ISO).

toxicants (Chapter 13) or between organisms (Chapter 16). Knowledge of both is required for full understanding of how a contaminated ecosystem functions. This is especially important as several of the sublethal effects of toxicants affect organisms by influencing their interactions with other organisms (section 16.5). Such interactions can be studied experimentally using micro- and mesocosms. These are completely defined entities that enable organismic interactions to be studied experimentally, thus forming a continuum from pure laboratory experiments to ecosystem monitoring. The principles of micro- and mesocosms are described in Figure 5.1. The differences between the two are mainly in size and complexity. Microcosms are typically flasks containing only a couple of species from the different trophic levels, lacking any of the large organisms of the food web. They are used, for example, when interactions between phytoplankton and zooplankton are studied. Mesocosms are more complex systems with several trophic levels. However, even in the most complex mesocosms, the long-living and large organisms (mostly predators) are lacking. Thus, if such organisms play an important role in the function of ecosystem, the information given by a mesocosm is limited. Another problem is that the equilibrium state reached in the mesocosm may be quite different from the equilibrium state of natural ecosystems, whereby the information obtained about the effects of environmental toxicants may be different from in natural cases. Another problem is that any experimentation with mesocosms has a limited time frame, which may cause toxicant responses to be different from what is observed in natural ecosystems. Difficulties

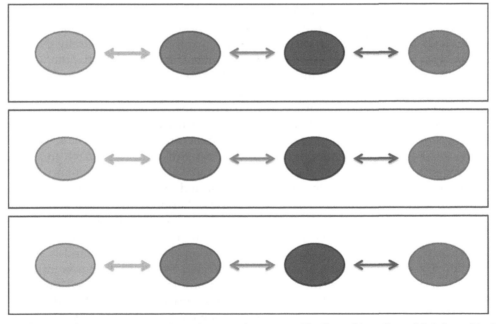

FIGURE 5.1 **A schematic representation of a micro-/mesocosm.** The "cosm" is replicated (triplicated in the figure). The several trophic levels interact. In the figure, the food web consists of phytoplankton (green ovals), zooplankton (blue ovals), primary predators (red ovals, e.g. crustaceans eating zooplankton), and secondary predators (orange ovals, e.g. small fish eating crustaceans). Often, mesocosms are artificial streams, which also include macroscopic green plants.

associated with these time limitations are that the community of the mesocosm may not have reached equilibrium and that the responses to a toxicant may be season-dependent, so that mesocosm studies often give results that are different according to the season in which the mesocosm experiment is carried out.

There are two major alternatives for generating the communities in the micro- and meso-cosms. The first is to use organisms from laboratory cultures, to get an exact standardization of the system. The number of seeded organisms will also be exactly defined. This guarantees that the replicates used in the controls and treatment communities are initially alike. However, the communities will not be like natural communities in the area. In the second alternative, the organisms seeded in the micro-/mesocosm are taken from natural settings. The community they form should then be allowed to reach equilibrium before the treatment starts. Although this alternative is ecologically relevant, it is very difficult or even impossible to define the organisms present in the seeds completely, and the replicates and their equilibrium communities necessarily deviate from each other. Also, the equilibrium reached may be different from the one that would have been reached in the natural ecosystem. Further, the development of equilibrium takes time, which may be the limiting resource in experimentation.

5.3 ECOSYSTEM MANIPULATIONS

Manipulations of full ecosystems have been carried out especially in freshwater systems. Complete manipulations of ecosystems have been done, for example, by dividing small lakes in two, with the two parts having different treatment, one side usually being the control and the other side receiving treatment. The rationale of these manipulations is naturally that the conditions in the system are the same apart from the treatment. This requires that the water quality remains the same on both sides of the division in the absence of treatment. Neither the inflows nor outflows of water in the compartments can be dissimilar. Furthermore, the bottom properties of both divisions must be the same. A successful partitioning of a pond also requires that no leakage of water from one partition to another occurs. Successful examples of partitioning lakes have been experiments where acidification and the effects of neutralization with chalk were studied.

Among all the ecosystem manipulations, the Experimental Lakes Area in Canada has been the most versatile. Because of its large size and versatility, its maintenance is also quite expensive, and it is currently under threat of being shut down to achieve short-term cost savings. However, the Experimental Lakes Area is the only one in the world where long-term consequences of environmental pollution can be followed in a relatively well-known setting and with replication of similar lakes. As the Experimental Lakes Area is unique in the world, its shutdown would be a great pity. The area was founded in 1967–68, and is situated in a sparsely inhabited region. Thus, it is largely unaffected by external human influences and industrial activities, and the results obtained from any treatments carried out necessarily describe the effects of those treatments and not some unwanted external influence.

Smaller-scale manipulations of total ecosystems have been carried out elsewhere in the world. One of the major problems in aquatic toxicology and aquatic biology in general is that the water properties of every system are unique, depending on the soil and mineral composition of the catchment area. Many of these properties affect the bioavailability of xenobiotics. Quite often this means that, even if experiments are replicated, the replications are not truly independent. For example, if an animal is living in free water, and the water properties (and not other properties of the aquatic ecosystem such as the vegetation, bottom sediments, etc.) are the major factors affecting its success, it is difficult to envision what additional information one gets from dividing the studied organisms into several entities, apart from the probability that their vulnerability is greater if they are all in one aquarium than if they are divided in several aquaria. The possible influence of containers increases with increasing dependence of the organism on the vegetation and bottom sediment. The fact that the work can only give information on the effects of contamination in that particular water requires that the water quality in any aquatic toxicology investigation is described in detail so that it is possible later, when different types of water have been used with the same toxicant, to evaluate which water qualities affect the behavior of the toxicant.

5.4 BIOMONITORING

Biomonitoring is the systematic use of organisms to evaluate the changes caused in an ecosystem by environmental contamination. Biomonitoring addresses both the exposure to contaminants and the biological responses to them. Thus, it uses bioindicator species and biomarkers, as discussed in Chapter 12. In addition to biological responses, it is important to determine the concentrations of the probable most important contaminants, so that one can estimate the likely reasons for the observed organismic effects. Biomonitoring is closely associated with environmental risk assessment (ERA). With the help of biomonitoring using bioindicators and biomarkers, one hopes to get an indication of whether adverse effects may take place (see Figure 5.2). Such a possibility constitutes an environmental risk. ERA is also discussed in more detail in Chapter 12. Organismic responses must be measured both in clean (control) and in impacted environments. Apart from the contamination, the water properties of the clean and impacted sites should be similar. Unless this is the case, it is nearly impossible to say if the observed differences between control and impacted organisms are due to contamination or to differences in the water quality between the measurement sites. The organisms used in the determinations can be either local or brought from a clean environment (and reared in a clean environment). Both types of organism have their advantages

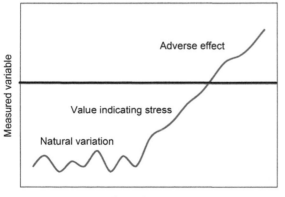

FIGURE 5.2 **The aims of biomonitoring.** The value of the measured parameter should show a change compared to natural variation in the parameter value. However, this stressed value, which indicates contaminant pressure, should precede any adverse effect so that the environmental risk of the contamination can be predicted before adverse effects take place.

and disadvantages, so it depends on the purpose of the study which is preferred. If one wants to evaluate long-term genetic responses, one needs to sample native organisms. However, if one is trying to evaluate what specific responses the environment is associated with in the short term, caged organisms must be used. The problem with native organisms is mainly that their age and previous exposure are usually difficult to define, whereas both can be accurately defined for caged organisms. This means that caged organisms enable one to determine both acute and delayed toxicant effects. On the other hand, any genetic responses are hard to define, especially for species with long generation times. As a consequence, the best possible alternative is to use both types of organisms to get as full a picture of environmental effects on organisms as possible.

One specific solution for the development of native communities for use in ecotoxicological research is to develop biofilms (short-life-cycle prokaryotes, microscopic algae, unicellular animals, etc.) on well-defined glass plates, and follow their development and the equilibrium reached in clean and contaminated environments for a defined period.

Relevant Literature and Cited References

Bailer, A.J., Oris, J.T., 2013. Aquatic toxicology. In: El-Shaarawi, A.H., Piegorsch, W.W. (Eds.), Encyclopedia of Environmetrics. Wiley, Hoboken NJ.

Bonada, N., Prat, N., Resh, V.H., Statzner, B., 2006. Developments in aquatic insect biomonitoring: A comparative analysis of recent approaches. Annu. Rev. Entomol. 51, 495–523.

Borja, A., Bricker, S.B., Dauer, D.M., Demetriades, N.T., Ferreira, J.G., Forbes, A.T., Hutchings, P., Jia, X., Kenchington, R., Carlos, M.J., Zhu, C., 2008. Overview of integrative tools and methods in assessing ecological integrity in estuarine and coastal systems worldwide. Mar. Poll. Bull. 56, 1519–1537.

Cairns Jr, J., Bidwell, J.R., Arnegard, M.E., 1996. Toxicity testing with communities: Microcosms, mesocosms, and whole-system manipulations. Rev. Environ. Contam. Toxicol. 147, 45–69.

Caquet, T., Lagadic, L., Sheffield, S.R., 2000. Mesocosms in ecotoxicology (I): Outdoor aquatic systems. Rev. Environ. Contam. Toxicol. 165, 1–38.

Colbourne, J.K., Pfrender, M.E., Gilbert, D., Thomas, W.K., Tucker, A., Oakley, T.H., Tokishita, S., Aerts, A., Arnold, G.J., Basu, M.K., Bauer, D.J., Caceres, C.E., Carmel, L., Casola, C., Choi, J.H., Detter, J.C., Dong, Q., Dusheyko, S., Eads, B.D., Frohlich, T., Geiler-Samerotte, K.A., Gerlach, D., Hatcher, P., Jogdeo, S., Krijgsveld, J., Kriventseva, E.V., Kultz, D., Laforsch, C., Lindquist, E., Lopez, J., Manak, J.R., Muller, J., Pangilinan, J., Patwardhan, R.P., Pitluck, S., Pritham, E.J., Rechtsteiner, A., Rho, M., Rogozin, I.B., Sakarya, O., Salamov, A., Schaack, S., Shapiro, H., Shiga, Y., Skalitzky, C., Smith, Z., Souvorov, A., Sung, W., Tang, Z., Tsuchiya, D., Tu, H., Vos, H., Wang, M., Wolf, Y.I., Yamagata, H., Yamada, T., Ye, Y., Shaw, J.R., Andrews, J., Crease, T.J., Tang, H., Lucas, S.M., Robertson, H.M., Bork, P., Koonin, E.V., Zdobnov, E.M., Grigoriev, I.V., Lynch, M., Boore, J.L., 2011. The ecoresponsive genome of *Daphnia pulex*. Science 331, 555–561.

Cold Spring Harbor Protocols. Emerging Model Organisms: A Laboratory Manual. http://cshprotocols.cshlp.org/site/emo/. Cold Spring Harbor Laboratory Press. Accessed 21 March 2013.

Doust, J.L., Schmidt, M., Doust, L.L., 1994. Biological assessment of aquatic pollution: A review, with emphasis on plants as biomonitors. Biol. Rev. Cambr. Phil. Soc. 69, 147–186.

Engeszer, R.E., Patterson, L.B., Rao, A.A., Parichy, D.M., 2007. Zebrafish in the wild: A review of natural history and new notes from the field. Zebrafish 4, 21–40.

Gerhardt, A., Ingram, M.K., Kang, I.J., Ulitzur, S., 2006. In situ on-line toxicity biomonitoring in water: Recent developments. Environ. Toxicol. Chem. 25, 2263–2271.

Guasch, H., Bonet, B., Bonnineau, C., Corcoll, N., Lopez-Doval, J.C., Munoz, I., Ricart, M., Serra, A., Clements, W., 2012. How to link field observations with causality: Field and experimental approaches linking chemical pollution with ecological alterations. Handbook Environ. Chem. 19, 181–218.

Kidd, K.A., Blanchfield, P.J., Mills, K.H., Palace, V.P., Evans, R.E., Lazorchak, J.M., Flick, R.W., 2007. Collapse of a fish population after exposure to a synthetic estrogen. Proc. Natl. Acad. Sci. USA 104, 8897–8901.

Lear, G., Lewis, G.D. (Eds.), 2012. Microbial Biofilms: Current Research and Applications. Caister Academic Press, Poole, UK.

Nikinmaa, M., Celander, M., Tjeerdema, R., 2012. Replication in aquatic biology: The result is often pseudoreplication. Aquat. Toxicol., 116–117. iii-iv.

Oikari, A., 2006. Caging techniques for field exposures of fish to chemical contaminants. Aquat. Toxicol. 78, 370–381.

Östlund-Nilsson, S., Mayer, I., Huntingford, F., 2007. The Biology of Three-Spined Stickleback. CRC Press, Boca Raton, FL.

Raisuddin, S., Kwok, K.W., Leung, K.M., Schlenk, D., Lee, J.S., 2007. The copepod *Tigriopus*: A promising marine model organism for ecotoxicology and environmental genomics. Aquat. Toxicol. 83, 161–173.

Zhou, Q., Zhang, J., Fu, J., Shi, J., Jiang, G., 2008. Biomonitoring: An appealing tool for assessment of metal pollution in the aquatic ecosystem. Anal. Chim. Acta 606, 135–150.

Factors Affecting the Bioavailability of Chemicals

Abstract

Bioavailability, uptake, metabolism, storage, and excretion of chemicals constitute toxicokinetics. Bioavailability is the potential for uptake of a substance by a living organism. It is usually expressed as the fraction that can be taken up by the organism in relation to the total amount of the substance available. In pharmacology, the bioavailability is the ratio of the amount of a compound in circulation after its extravenous application and its intravenous injection. In aquatic toxicology, environmental bioavailability is usually relevant. Factors affecting the bioavailability of a chemical depend on the route of uptake, and whether the chemical is in the bottom sediment, dissolved in water, or is a constituent of the organisms. In the case of water-soluble substances, the primary source of toxicant is water, and the bioavailability depends on complex formation, especially with humic substances. Even when water-soluble substances are sediment bound, they reside mainly in pore water. Lipid-soluble substances are taken up especially from sediment or from other organisms. The bioavailability from water decreases with increasing lipophilicity and with increasing amount of dissolved organic carbon or colloids in the aquatic phase. With regard to sediment, both the sediment properties (e.g. grain size) and the amount of organic material in the sediment affect bioavailability. The main abiotic factors affecting bioavailability are oxygenation and pH. As an example, metal speciation, affecting bioavailability, depends very much on the pH.

Keywords: toxicokinetics; pharmacological bioavailability; absorbed dose fraction; humus; humic acids; fulvic acids; sediment; metal speciation; total organic carbon; dissolved organic carbon; sorption; nanomaterial bioavailability.

An Introduction to Aquatic Toxicology
http://dx.doi.org/10.1016/B978-0-12-411574-3.00006-2

6.1 INTRODUCTION

The bioavailability of a toxicant, its uptake, metabolism (including transformation), storage, and excretion make up toxicokinetics. Toxicokinetics can thus be defined as the characteristics that affect the amounts of chemicals in the organism (Figure 6.1), and these constitute the material for this and the following chapters (Chapters 7, 8, 9, and 10). The first step of toxicokinetics is bioavailability. Bioavailability is the potential for uptake of a substance by a living organism. It is usually expressed as the fraction that can be taken up by an organism in relation to the total amount of the substance available. Notably, it is always the bioavailable fraction of a compound that participates in the uptake and consecutive responses to a chemical, not the total amount. Consequently, a smaller toxicant load in the environment can cause a larger response, if the environment is associated with increased bioavailability (Figure 6.2). Different aspects of bioavailability in the environmental context are given in Figure 6.3. The bioavailability can be divided into pharmacological bioavailability (how the route of administration affects the potential for uptake of a chemical) and environmental bioavailability (how interactions with the environment affect the potential for uptake of a chemical).

6.2 PHARMACOLOGICAL BIOAVAILABILITY

In pharmacology, the bioavailability of a compound is mainly determined by the route of administration. The bioavailability of a compound is virtually complete when the route of

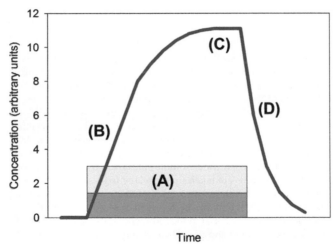

FIGURE 6.1 **Representation of different components of toxicokinetics as seen in the toxicant level in an organism as a function of time.** (A) An organism is exposed to a chemical. Only a portion of the chemical load is bioavailable (total concentration yellow + green, bioavailable fraction green). (B) The chemical is taken up by the organism. The initial uptake rate is a linear function of the bioavailable fraction of the chemical in the environment. However, after a time, the rate of uptake becomes curvilinear, as the chemical that is taken up is metabolized and excreted. (C) After a given (chemical-specific) time, a steady state is reached upon continuous exposure, where the uptake and metabolism/excretion are equal. (D) If exposure to the toxicant ends, its concentration in the organism decreases. The time course of depuration depends on the excretion mechanisms.

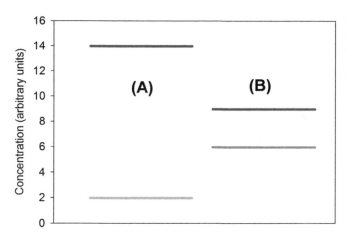

FIGURE 6.2 A hypothetical example showing that although the total concentration of a chemical is higher in environment A than in environment B, the bioavailable fraction in the latter (red line) can be greater than that in the former (green line).

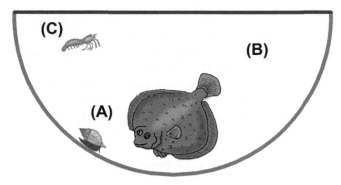

FIGURE 6.3 Bioavailability of chemicals by location in the environment. A chemical can be available for uptake by an organism if it is (A) contained in the sediment—this is the case especially for organic compounds and benthic organisms or organisms with roots in sediment; (B) in water—this is the case for water-soluble compounds such as most metal cations and small anions. Apparently water-bound organic compounds are often associated with dissolved organic matter or colloidal material in the water column. (C) Chemicals can also be available for uptake from food organisms—normally lipid-soluble compounds are bioavailable through ingestion of food such as prey organisms.

administration is intravenous injection, and varies with intraperitoneal injection, administration in food, administration in the respiratory medium, or penetration through the organismal surface. The bioavailability is given in those cases from the ratio of the systemic exposure from extravascular (e.v.) exposure to that following intravenous (i.v.) exposure, as described by the equation:

$$F = A_{ev} \, D_{iv} / B_{iv} \, D_{ev} \tag{6.1}$$

where F (absorbed dose fraction) is a measure of the bioavailability; A and B are the areas under the (plasma) concentration–time curve following e.v. and i.v. administration, respectively; and D_{ev} and D_{iv} are the administered e.v. and i.v. doses. This pharmacological definition of bioavailability is important in aquatic toxicology, as in experimental manipulations

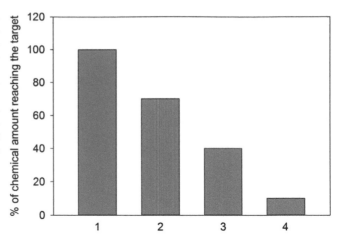

FIGURE 6.4 **Different routes of administration of a chemical leading to differences
in toxicity.** A hypothetical example of the
response to a toxicant given via (1) intravenous injection, (2) intraperitoneal injection,
(3) force feeding (gavage), or (4) exposure
to the chemical in the natural environment.
Although the molar amount of the chemical that the organism is exposed to can be
the same, the amount reaching the target
structure may be very different because
of the factors affecting (pharmacological)
bioavailability.

different toxicants can be administered via intravenous/intraperitoneal injections, via gavage
(force feeding), or as a part of the normal diet, or the organisms may be exposed to toxicants
via water. It should be remembered that whenever intravenous/intraperitoneal/subcutane-
ous injections are used, the normal uptake of chemicals from the environment (see Chapter 7)
is bypassed. Thus, any such studies can give information about the toxic actions of the compound when in the animal, but the findings cannot be considered environmentally relevant
as the possibilities and mechanisms of uptake have not been considered (see Figure 6.4 for a
schematic representation of differences in toxicity due to different ways of administering the
chemical).

Factors that affect the appearance of a compound (drug) in the systemic circulation affect
its pharmacological bioavailability. Such factors include interactions with other drugs or food
components taken concurrently (altering absorption), first-pass metabolism, intestinal motility, chemical degradation of the drug by intestinal microflora, physical properties of the drug
(hydrophobicity, pKa, solubility) and its formulation (including the speed and duration of the
release of the active ingredient from the encapsulation), whether the formulation is administered to a fed or fasted organism, rate of gastric emptying, circadian differences, age, and
gender.

6.3 ENVIRONMENTAL BIOAVAILABILITY

There are three major environmental components from which chemicals may enter organisms in the aquatic environment: water, bottom sediment, and other organisms. The importance of these as uptake sources varies depending on the water solubility of the compound.
Highly water soluble compounds are preferentially taken up from the aqueous phase. Even
if the compounds are largely found in sediment, their location and uptake site will be in the
pore water. Lipophilic compounds will be obtained either from sediment or, especially, from
food (i.e. other organisms). The deposition history of a compound will also affect its availability. If most of the compound has accumulated to bottom sediments over time, the potential for

uptake from these sediments will be highest. Finally, different organisms can have markedly different loads of contaminants depending on their feeding habits, their movements between differently contaminated areas, etc. As an example, a sediment-bound contaminant can be transferred in the food web quite differently in the presence of rooted green plants than in the presence of pelagic unicellular algae. Regardless of the main uptake route of the chemical, the factors affecting bioavailability are essentially the same. They are, first, the complex formation of the toxicant with naturally occurring compounds; second, their lipophilicity; and, third, their resemblance to compounds that are specifically taken up. Naturally, the stability of the compound, which may be different in different compartments of aquatic systems, will also influence how it can be taken up. Strictly speaking, this is not an aspect of bioavailability, but a part of the environmental fate of the toxicant. However, the interactions of the chemical with the environment are an important determinant of the subsequent uptake, and therefore must be mentioned at this point.

In the bioavailability of compounds one needs to consider both the water and the sediment properties, and also how other organisms affect the properties of compounds taken up. An important component of such organismic interactions is that microorganisms, in particular, may take up substances and render them inaccessible to other organisms that do not eat the microorganisms. Also, when considering bioavailability, the delivery of compounds from the atmosphere by rainfall (frequency is more important than amount) must be taken into account.

The total organic carbon content (TOC) and dissolved organic carbon content (DOC) have intimate association with bioavailability. Largely, this is due to many contaminants forming complexes with the carbon compounds. In aquatic systems the most important complex-forming compounds are humic substances. Humic substances are usually large colloidal molecules with several carboxylic acid and phenolic groups. They are usually divided into two classes: the larger humic acids (molecular weight (MW) > 1000, up to 100,000) and the smaller fulvic acids (MW usually < 1000). Because the humic substances contain carboxylate groups, they are normally acidic with an overall negative charge. In solution they usually behave as biphasic weak acids with a pK value around 4. The humic substances are formed by the breakdown of plant matter, and their exact structure is site-dependent, depending on the plants in the environment. Humic substances can also form a major part of the nutrition of prokaryotes. As the humic substances have an overall negative charge, they can form complexes with positively charged metal ions (and other positively charged compounds). Complex formation is especially pronounced with divalent cations, such as Fe^{2+}, Mg^{2+}, and Ca^{2+}.

The bioavailability of metals has been studied very intensively. In addition to complex formation, especially with humus, this is affected by the type of metal compound. For example, sulfide minerals may be encapsulated in quartz or other chemically inert minerals, and despite high total concentrations of metals in sediment containing these minerals they may not be bioavailable and thus their environmental effects may remain small. Consequently, as an example, the type of ore containing the metal affects the toxicity of mining effluents to aquatic systems. If the aquatic environment has reducing conditions, for example if they are hypoxic, metal ions are associated with sulfides, e.g. insoluble FeS is formed. Most metal sulfides are poorly soluble, and consequently quite immobile as long as they remain in a chemically reducing environment. Because of this, their bioavailability is reasonably low. Consequently, oxygenation of the aqueous medium (and the sediment) will affect the bioavailability of metals. The reason why arsenic has become a major pollutant is that it is released from the sulfide-based iron ores when they come into contact with oxygen-containing water.

In addition to affecting the fate of metals, oxygenation affects both the properties and stability of organic contaminants. The apparent presence and bioavailability of organic toxicants is different in oxic and anoxic sediments. Another sediment property affecting the bioavailability of organic contaminants is the amount of other organic material in the sediment. Increasing organic material tends to increase the sorption of an organic chemical to the sediment, reducing its availability in the aquatic phase. In general, aquatic bioavailability has close interactions with how compounds are moving between bottom sediments and water, and the final equilibria between the two compartments. Particle size and resulting total surface area available for adsorption are both important factors in adsorption processes in the sediments, and can affect the bioavailability of compounds. Small particles with large surface-area-to-mass ratios allow more adsorption than an equivalent mass of large particles with small surface-area-to-mass ratios. This is most important with regard to nanomaterials. Because of the surface-area-to-mass dependence of bioavailability, the toxicity of nanomaterials can be strongly affected by the size of particles in the formulations. Since the sedimentation of nanomaterials is affected by the way a test is executed, markedly different bioavailabilities and toxicities of nanomaterials have been reported. Reduced adsorption of compounds to sediments can also affect their bioavailability by affecting the dissolved concentration of a compound in the water surrounding the sediment.

In the aqueous phase, the bioavailability of organic contaminants depends on their partitioning between the organic phase and water. Lipophilicity is also an important aspect of chemical uptake in organisms, and its determination and main discussion are given in Chapter 7. However, it needs to be pointed out here that, because of the interactions between organic contaminants and other organic materials in water, the bioavailability of organic contaminants from water will decrease with increasing dissolved organic carbon content and increasing amount of colloids in the aquatic phase.

Apart from oxygenation, another abiotic factor affecting bioavailability is pH. The effects of pH have been studied especially with regard to the speciation (and solubility) of metals, as exemplified for aluminum in (Figure 6.5). At pH values above neutral, aluminum ions form insoluble compounds; between pH values 5 and 7, the metal exists

FIGURE 6.5 **Approximate proportions of the different aluminum species (free cations, hydroxides) as a function of pH.**

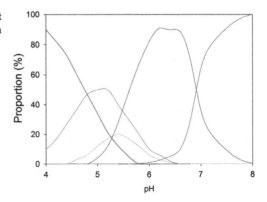

largely as soluble hydroxides; and at pH values below 5, it exists largely as toxic free Al^{3+} ions. Because of the pH effects, the bioavailability of metals and other compounds may be markedly different in bulk water and in the vicinity of (fish) gills, since there is a large pH gradient in the water from the gills to the bulk water. Also, the mode of toxicity of metals and other compounds may be different at different pH values. Again, as an example, aluminum appears to be mainly a respiratory toxicant at intermediary pH values (5–7), largely because it precipitates on gill filaments, increasing the diffusion distance of oxygen. At lower pH values, it appears to be an ionoregulatory toxicant. The hardness of water (essentially its salinity) also affects bioavailability, mostly of ionic compounds. The effect is largely due to a change in the probability of a toxic ion being taken up instead of a nontoxic one in the uptake site.

The bioavailability of metals is often calculated using the biotic ligand model (BLM), which seeks to take into account the effects of complex formation, abiotic environmental factors such as pH, and metal interactions (for details, see Chapter 18). Many of these factors vary seasonally and temporally, and most factors are interrelated. Consequently, changing one factor may affect several others. In addition, generally poorly understood biological factors seem to influence bioaccumulation of metals and thereby affect any predictions of their toxicity: an important factor here is the type of aquatic environment. Uptake of ions occurs very differently in marine and freshwater environments. Consequently, what is a useful model in freshwater may not be accurate in the marine environment. The modeling of toxicant accumulation in organisms is discussed further in Chapter 18.

Relevant Literature and Cited References

Andrady, A.L., 2011. Microplastics in the marine environment. Mar. Pollut. Bull. 62, 1596–1605.

Beckett, R., Jue, Z., Giddings, J.C., 1987. Determination of molecular weight distributions of fulvic and humic acids using flow field–flow fractionation. Environ. Sci. Technol. 21, 289–295.

Boudou, A., Ribeyre, F., 1997. Aquatic ecotoxicology: From the ecosystem to the cellular and molecular levels. Environ. Health Perspect. 105 (Suppl. 1), 21–35.

Burton Jr., G.A., 2010. Metal bioavailability and toxicity in sediment. Crit. Rev. Environ. Sci. Technol. 40, 852–907.

Camargo, J.A., 2003. Fluoride toxicity to aquatic organisms: A review. Chemosphere 50, 251–264.

Chapman, P.M., Wang, F., Janssen, C., Persoone, G., Allen, H.E., 1998. Ecotoxicology of metals in aquatic sediments: Binding and release, bioavailability, risk assessment, and remediation. Can. J. Fish Aquat. Sci. 55, 2221–2243.

Farrington, J.W., 1991. Biogeochemical processes governing exposure and uptake of organic pollutant compounds in aquatic organisms. Environ. Health Perspect. 90, 75–84.

Haitzer, M., Hoss, S., Traunspurger, W., Steinberg, C., 1998. Effects of dissolved organic matter (DOM) on the bioconcentration of organic chemicals in aquatic organisms: A review. Chemosphere 37, 1335–1362.

Hamelink, J.L., Landrum, P.F., Bergman, H.L., Benson, W.H., 1994. Bioavailability: Physical, Chemical and Biological Interactions. CRC Press, Boca Raton, FL.

Haws, N.W., Ball, W.P., Bouwer, E.J., 2006. Modeling and interpreting bioavailability of organic contaminant mixtures in subsurface environments. J. Contam. Hydrol. 82, 255–292.

Henry, T.B., Petersen, E.J., Compton, R.N., 2011. Aqueous fullerene aggregates (nC60) generate minimal reactive oxygen species and are of low toxicity in fish: A revision of previous reports. Curr. Opin. Biotechnol. 22, 533–537.

Markich, S.J., 2002. Uranium speciation and bioavailability in aquatic systems: An overview. Sci. World J. 2, 707–729.

Navarro, E., Baun, A., Behra, R., Hartmann, N.B., Filser, J., Miao, A.J., Quigg, A., Santschi, P.H., Sigg, L., 2008. Environmental behavior and ecotoxicity of engineered nanoparticles to algae, plants, and fungi. Ecotoxicology 17, 372–386.

Porcal, P., Koprivnjak, J.F., Molot, L.A., Dillon, P.J., 2009. Humic substances—part 7: The biogeochemistry of dissolved organic carbon and its interactions with climate change. Environ. Sci. Pollut. Res. Int. 16, 714–726.

Rudel, H., 2003. Case study: Bioavailability of tin and tin compounds. Ecotoxicol. Environ. Saf. 56, 180–189.

Schrap, S.M., 1991. Bioavailability of organic chemicals in the aquatic environment. Comp. Biochem. Physiol. C 100, 13–16.

Sharma, V.K., 2009. Aggregation and toxicity of titanium dioxide nanoparticles in aquatic environment: A review. J. Environ. Sci. Health A 44, 1485–1495.

Tessier, A., Turner, D.R., 1995. Metal Speciation and Bioavailability in Aquatic Systems. Wiley, Hoboken, NJ.

Worms, I., Simon, D.F., Hassler, C.S., Wilkinson, K.J., 2006. Bioavailability of trace metals to aquatic microorganisms: Importance of chemical, biological and physical processes on biouptake. Biochimie 88, 1721–1731.

Chemical Uptake by Organisms

Abstract

Factors affecting the uptake of chemicals are discussed. The uptake is usually measured via measuring the fluxes of radioactive compounds. Fluxes of organic compounds are usually measured with C-14- or H-3-labeled compounds, because they have low energy of radioactive emission. For metals (and inorganic compounds), a radioactive isotope of the metal (or inorganic anion) in question is usually used for flux measurements. As a generalization, in animals with gills and intestine, hydrophilic compounds are taken up by gills in freshwater, and mostly gills but also intestine in the marine environment, where animals drink to obtain the water needed. If organisms do not possess these richly vascularized thin epithelia, uptake occurs through the general surface. Hydrophilic contaminants reach the uptake sites easily in the aqueous medium, but rapid uptake requires the presence of carriers for which the contaminant has a significant affinity. In contrast, lipophilic compounds are poorly transported in aqueous media, but after reaching the vicinity of uptake sites, are effectively taken up. Their primary sources are sediments and food organisms. The lipophilic (organic) contaminants bioaccumulate in the organism and biomagnify along the trophic chain.

Keywords: dose; bioaccumulation; biomagnification; flux measurement; octanol/water partition coefficient; hydrophilic compound; lipophilic compound; gill uptake; intestinal uptake; aqueous uptake; uptake from sediment; uptake from food.

7.1 INTRODUCTION

To have toxic effect, a substance has to get into the organism, i.e. be bioavailable (Chapter 6). In the case of pelagic organisms, only the contaminants dissolved in water or those contained in food can be taken up. For organisms with roots, or bottom-dwelling

73

ones, the sediment can also be important as the source of uptake. The uptake efficiency depends on the uptake sites. Because the principles of uptake are quite different for water- and lipid-soluble compounds, the two are discussed in separate sections below. Regardless of the uptake route, a distinction between the amount of chemical entering the target (the dose) and the amount of chemical in the environment (often the concentration in water, but it can be also the amount of chemical in a given volume of sediment or in a given weight of food) must be made. Surprisingly often, studies in aquatic toxicology equate the dose and water concentration. However, the two can be markedly different: if a chemical is not taken up by the animal, the dose is negligible even if the chemical concentration in the environment is high. Even if the compound is highly toxic when injected to an organism, a limited uptake may result in apparently limited toxicity. Because of the fact that the environmental level and the amount in an organism may be markedly different, and because dose refers to the amount of material reaching the target organism, it is a mistake to say "dose" or "dose dependency" in the case where the amount of chemical in the organism is not measured, but only the concentration in the water surrounding the organism. In such cases, "concentration" or "environmental concentration" and "concentration dependency" should be used instead. The pronounced use of "dose" in aquatic toxicology even when this usage is faulty probably stems from general (mammalian) toxicology where a major proportion of studies use "injected doses," where the use of dose is correct as the word refers to the amount getting into the animal. The situation is similar for the cases where chemicals are injected directly into aquatic animals. However, as discussed in section 6.2, chemical uptake from the environment and its regulation are then bypassed. Another problem, which has significant bearing on aquatic toxicological studies in general and uptake studies in particular, is that occasionally nominal concentrations instead of measured ones are given. Increasingly, this habit is not accepted by journals, as large variations between nominal and actual concentrations are possible. This is the case if a compound is degraded, adsorbed to the surfaces in the container, or evaporates to the atmosphere. Naturally, measurements of chemical concentrations are not always possible, but in those cases the use of nominal concentrations in the study must be justified. Undoubtedly the large variation often seen in reported toxicities of a compound in one species is partially due to the limited knowledge of the actual concentration of the chemical in the effect site (due to limited knowledge of both the actual concentration and the uptake to the organism).

The uptake of a compound is usually measured with radioactively labeled isotopes. For most organic compounds, either C-14- or H-3-labeled material is used. C-14 and H-3 are used because the energy of their beta radiation is low, whereby the radiotoxicity of the chemical is small. For metal-ion uptake, the radioactive ion (e.g. Na-24 for sodium) is the measured entity. Rubidium-96 is often used instead of potassium. Even though the molecular weight of rubidium is more than twice that of potassium, the isotope is considered to behave qualitatively similarly to potassium. One of the reasons for this is that the hydrated radius of the rubidium ion is similar to that of the potassium ion, and it is the hydrated ion that is thought to be transported. The reason why radioactive potassium ions are not often used is that their half-life is very short, maximally much less than a day. Consequently, the isotope should be prepared in the vicinity of where the uptake experiments are carried out. Uptake measurements, the principles of which are given in Figure 7.1, show a straight line

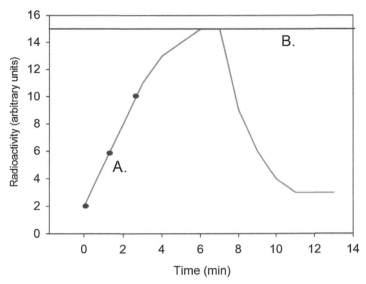

FIGURE 7.1 **Principles of flux measurements.** The radioactive isotope is added to the water (extracellular compartment). Thus, at time 0 there is no radioactivity in the intracellular compartment, and, consequently, any radioactivity measured indicates the size of the extracellular compartment. Its size as a % can be quantified from the 0 time-value of radioactivity, but normally impermeable extracellular markers such as inulin or polyglycol are used. Uptake/influx can be estimated from the appearance of the radioactivity in the intracellular (organismic) compartment. For the estimation of influx, first, the specific activity of the radioactive isotope in the water (extracellular compartment) must be known (becquerels per millimole contained in unit weight); second, in comparison to the organismic compartment, the water compartment should be infinite so that uptake does not cause significant depletion of extracellular radioactivity; and, third, minimally two (in addition to the 0-time measurement) measurements (measurements given as red circles) of intracellular radioactivity should be available during the time period where the increase of internal radioactivity is linear. In this case, the flux in moles per unit time can be estimated from the linear portion of the radioactivity–time curve. The steady-state level (depicted with a blue line) indicates bioaccumulation.

at the beginning. The slope gives the unidirectional uptake rate, which is often the property of interest. Later, there is both in- and efflux, and when they are equal, the system is in steady state.

Regardless of the type of toxicant, the uptake in animals occurs mainly via the thinnest epithelia, i.e. the gills and the gut. These tissues are also richly vascularized, whereby the contaminants are effectively removed from the uptake site to the systemic targets. In organisms without gills or gut, uptake occurs through the general surface of the organism. In animals with an exoskeleton, very little uptake occurs via the general body surface. Although chemical uptake via the general body surface is also small in fish, because the diffusion distance from the environment to the circulation is long, some chemical uptake occurs.

The route of uptake is usually associated with the lipophilicity/hydrophilicity of the compound. The lipophilicity is given by the octanol/water partition coefficient, K_{ow}. It is traditionally determined by shaking the compound in an octanol/water mixture, allowing the octanol and water to separate, and measuring the concentration of the compound in each.

FIGURE 7.2 **A schematic representation of metal uptake in fish gills.** (A) The uptake occurs mainly via carriers with preference for their natural substrates. However, the carriers can often also transport other metals, albeit with much lower affinity. (B) Slower metal uptake can occur via the paracellular route or (C) across the cell membrane. The highest densities of metal carriers are present in chloride cells, situated especially at the base of secondary lamellae. The figure shows a schematic representation of a secondary lamellum with blood spaces (erythrocytes depicted as red ovals), cell borders, and metal carriers (depicted as rounded blue rectangles).

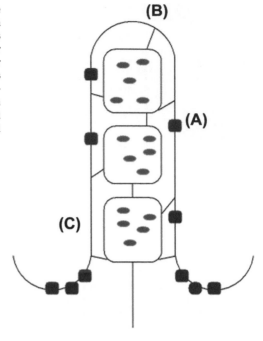

The higher the ratio of compound in octanol/compound in water, the higher is its lipophilicity. Often, the lipophilicity is given as P_{Kow}, which gives the logarithm ($\log_{10}$) of the ratio. In animals, lipophilic compounds are mainly taken up in food and enter the body in the intestine. Hydrophilic compounds, such as metal ions, are taken up mainly via gills, although a contribution from the gut is important in marine animals. To a small extent, hydrophilic compounds are also taken up by the general body surface.

7.2 THE UPTAKE OF IONIC (HYDROPHILIC) COMPOUNDS

For hydrophilic compounds, the molecule is easily transported to the vicinity of the uptake site, but its transport through the lipid membranes is very slow unless facilitated by protein carriers or pores. As an example, transport of chloride across the membrane of lamprey erythrocytes, devoid of anion carriers, has a half-time of about two hours, but that in teleost erythrocytes only seconds. The speed of uptake of hydrophilic toxicants thus depends on the availability and affinity of carriers transporting the toxicant across the lipid barriers. Toxicants will be transported by carriers that are primarily aimed at transporting chemicals required by the organisms. In particular, the uptake of toxic metals has been studied in detail. The principles of metal uptake by gill epithelial cells are given in Figure 7.2. The metal uptake may occur via the sodium and calcium pumps, by the sodium/potassium/chloride cotransporter, or by the sodium/proton exchanger. Although these are the more common transporters, present in most cells, small numbers of carriers for the ions of iron, etc., can be found. With regard to anions, different types

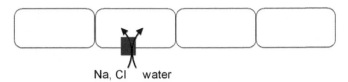

Na, Cl   water

FIGURE 7.3   **A schematic representation of metal and water uptake in the intestine in seawater.** Metal ions are taken up via specific carriers (depicted as a blue rectangle) and osmotically obliged water follows. The salt (metal—especially sodium—and anion—especially chloride) load is actively extruded in gills.

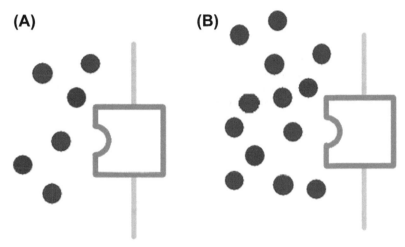

FIGURE 7.4   **Effect of salinity on metal uptake.** (A) The probability of toxic metal (blue symbols) uptake depends on the affinity of the carrier for its normal substrate (red symbols) and the toxic metal. (B) With increasing salinity, the concentration of the normal metal ion substrate increases, which decreases the probability of the toxic metal uptake.

of anion exchangers, some carrying out cotransport of an anion and sodium or protons, are found. Small organic anions can be transported through carboxylic acid and glucose transporters. In freshwater animals, gills are the sole route for hydrophilic toxicant transport. Freshwater animals do not drink, as they try to avoid obtaining excess water. On the other hand, seawater fish do drink, and thereby the uptake of hydrophilic compounds also occurs in the intestine. The principles of intestinal water and ion uptake are given in Figure 7.3. Because the metal affinities of transporters in different tissues vary, the metal uptake will change with a change in the proportion of uptake occurring in the intestine. Even with the metal affinity of the transport remaining constant, the rate of metal uptake is an inverse function of salinity, since the probability of a metal ion entering the transport site decreases with increasing salinity, as described in Figure 7.4.

Complex formation of toxic metals with, for example, humic acids decreases the probability of their being transported to the gill and decreases transport across lipid membranes when the ion is bound to the carrier, but increases transport directly across the lipid barrier, because the lipid solubility of the complex is higher than that of the ion alone. The question of uptake mechanism is very relevant for metal-containing nanomaterials, because differentiating between the toxicity of metal ions liberated from them and that of the nanomaterial itself is an issue.

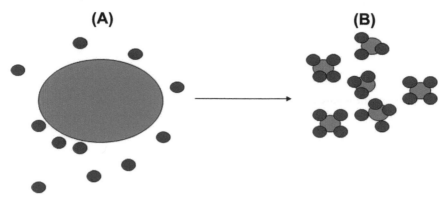

FIGURE 7.5   **Micelle formation from lipid droplets as a result of bile acid function.** (A) Detergents such as bile acids surround a lipid droplet and decrease its surface tension, whereby (B) the droplet is broken down into small micelles, which are able to come close to the intestinal walls, and consequently the lipid-soluble substances in the micelles are able to cross the lipid membranes of the cells in the intestinal wall.

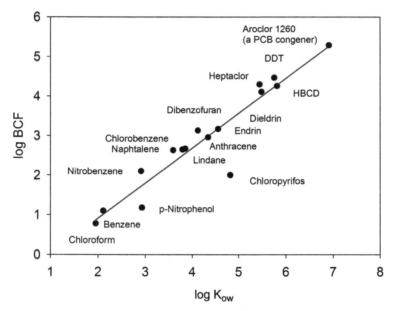

FIGURE 7.6   **The bioconcentration and lipophilicity (given as P_{Kow} = logarithm of octanol/water distribution ratio, log K_{ow}) of some environmental contaminants.** HBCD, hexabromocyclododecane; PCB, polychlorinated biphenyl.

7.3 THE UPTAKE OF LIPOPHILIC COMPOUNDS

Lipophilic compounds can cross any lipid barrier easily, but cannot easily reach the barrier. Consequently, they are hardly ever taken up from the water via the gills in animals. In contrast, they are taken up from sediments and food organisms in the intestine. The probability of coming into contact with the lipid membranes of cells lining the intestine is increased

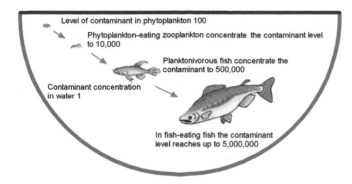

Level of contaminant in phytoplankton 100

Phytoplankton-eating zooplankton concentrate the contaminant level to 10,000

Planktonivorous fish concentrate the contaminant to 500,000

Contaminant concentration in water 1

In fish-eating fish the contaminant level reaches up to 5,000,000

FIGURE 7.7 **Biomagnification of a lipophilic environmental contaminant across trophic levels.**

by the detergent action of bile acids, which transform any lipid droplets into small micelles (Figure 7.5).

The lipophilicity of a compound also affects its probability of bioaccumulation/bioconcentration. Figure 7.6 gives the correlation between the bioconcentration factor and lipophilicity of some environmental contaminants. Once a lipophilic compound is taken up by an organism, it is retained in its lipid, and not lost to any aqueous media. Consequently, lipophilic compounds can be bioaccumulated so that their measured content in an animal is in some cases many thousands of times higher than that in the same amount of aqueous solution. Bioaccumulation refers to an increase of accumulated level in a single organism. Biomagnification refers to an increase in the amount of chemical across trophic levels. The more lipophilic a compound is, the more likely its concentration is to increase in the different trophic levels in the food chain from phytoplankton to top predators such as seals (Figure 7.7). Good examples of compounds with pronounced biomagnification are organochlorine insecticides such as DDT, which can be up to 100,000,000 times more concentrated in fish-eating birds than in the water. Quite often, bioaccumulation is thought to decrease at very high lipophilicities (pK > 5.8). However, this bioaccumulation cutoff may be a result of problems in experimentation with highly lipophilic compounds. It is important to note that the bioaccumulation of lipophilic compounds depends on the temperature, presence of humic acids, and pH. If a lipophilic compound is a weak acid, pH affects the proportions of acid (uncharged) and base (charged) forms. Of these, the acid form is much more permeable. Consequently, the uptake will be markedly reduced with an increase in pH.

Relevant Literature and Cited References

Ashauer, R., Caravatti, I., Hintermeister, A., Escher, B.I., 2010. Bioaccumulation kinetics of organic xenobiotic pollutants in the freshwater invertebrate *Gammarus pulex* modeled with prediction intervals. Environ. Toxicol. Chem. 29, 1625–1636.

Baun, A., Sorensen, S.N., Rasmussen, R.F., Hartmann, N.B., Koch, C.B., 2008. Toxicity and bioaccumulation of xenobiotic organic compounds in the presence of aqueous suspensions of aggregates of nano-C(60). Aquat. Toxicol. 86, 379–387.

Bury, N., Grosell, M., 2003. Iron acquisition by teleost fish. Comp. Biochem. Physiol. C 135, 97–105.

Connell, D.W., 1989. Bioaccumulation of Xenobiotic Compounds. CRC Press, Boca Raton, FL.

Ellory, J.C., 1982. Flux measurements. In: Techniques in Cellular Physiology, Pt. II. Elsevier/North-Holland, New York, pp. 1–11.

Fent, K., Looser, P.W., 1995. Bioaccumulation and bioavailability of tributyltin chloride: Influence of pH and humic acids. Water Res. 29, 1631–1639.

Haitzer, M., Hoss, S., Traunspurger, W., Steinberg, C., 1998. Effects of dissolved organic matter (DOM) on the bioconcentration of organic chemicals in aquatic organisms: A review. Chemosphere 37, 1335–1362.

Hendriks, A.J., Heikens, A., 2001. The power of size, 2: Rate constants and equilibrium ratios for accumulation of inorganic substances related to species weight. Environ. Toxicol. Chem. 20, 1421–1437.

Jonker, M.T., van der Heijden, S.A., 2007. Bioconcentration factor hydrophobicity cutoff: An artificial phenomenon reconstructed. Environ. Sci. Technol. 41, 7363–7369.

Katagi, T., 2010. Bioconcentration, bioaccumulation, and metabolism of pesticides in aquatic organisms. Rev. Environ. Contam. Toxicol. 204, 1–132.

Landner, L. (Ed.), 1989. Chemicals in the Aquatic Environment: Advanced Hazard Assessment. Springer, Berlin.

Lundqvist, A., Bertilsson, S., Goedkoop, W., 2012. Interactions with DOM and biofilms affect the fate and bioavailability of insecticides to invertebrate grazers. Ecotoxicology 21, 2398–2408.

Mackay, D., 1982. Correlation of bioconcentration factors. Environ. Sci. Technol. 16, 274–278.

Mason, R.P., 2013. Trace Metals in Aquatic Systems. Wiley-Blackwell, Oxford, UK.

McFarlane, C., Trapp, S. (Eds.), 1994. Plant Contamination: Modeling and Simulation of Organic Chemical Processes. CRC Press, Boca Raton, FL.

Monteiro, C.M., Castro, P.M., Malcata, F.X., 2012. Metal uptake by microalgae: Underlying mechanisms and practical applications. Biotechnol. Prog. 28, 299–311.

Muncaster, B.W., Hebert, P.D., Lazar, R., 1990. Biological and physical factors affecting the body burden of organic contaminants in freshwater mussels. Arch. Environ. Contam. Toxicol. 19, 25–34.

Playle, R.C., Wood, C.M., 1989. Water chemistry changes in the gill micro-environment: Experimental observations and theory. J. Comp. Physiol. B 159, 527–537.

Playle, R.C., Wood, C.M., 1989. Water pH and aluminum chemistry in the gill micro-environment of rainbow trout during acid and aluminum exposures. J. Comp. Physiol. B 159, 539–550.

Rainbow, P.S., 2007. Trace metal bioaccumulation: Models, metabolic availability and toxicity. Environ. Int. 33, 576–582.

Sangster, J., 1989. Octanol–water partition coefficients of simple organic compounds. J. Phys. Chem. Ref. Data 18, 1111–1227.

Smith, D.J., Gingerich, W.H., Beconi-Barker, M.G. (Eds.), 1999. Xenobiotics in Fish. Springer, New York.

Veltman, K., Huijbregts, M.A.J., van Kolck, M., Wen-Xiong Wang, W.-X., Hendriks, A.J., 2008. Metal bioaccumulation in aquatic species: Quantification of uptake and elimination rate constants using physicochemical properties of metals and physiological characteristics of species. Environ. Sci. Technol. 42, 852–858.

Wang, W.X., Rainbow, P.S., 2010. Significance of metallothioneins in metal accumulation kinetics in marine animals. Comp. Biochem. Physiol. C 152, 1–8.

Wood, C.M., 1992. Flux measurements as indices of proton and metal effects on freshwater fish. Aquat. Toxicol. 22, 239–264.

Wood, C.M., Playle, R.C., Hogstrand, C., 1999. Physiology and modeling of mechanisms of silver uptake and toxicity in fish. Environ. Toxicol. Chem. 18, 71–83.

8

Chemical Distribution in Organisms

Abstract

Once in an organism, chemicals are either metabolized or stored. Metabolism includes both toxic effects and detoxification. In addition to metabolism, storing chemicals in inert material can be considered a detoxification mechanism. On the other hand, release of toxic chemicals from inert material can cause delayed toxicity. The route of uptake may affect the distribution and effects of contaminants entering the organism. In contrast to the whole-body distribution, which has functional ramifications, the subcellular distribution of chemicals is often defined operationally, and the same operational compartment can include very different functional components.

Keywords: storage; delayed toxicity; exoskeleton; bone; shell dissolution; fat tissue; toxicant distribution; vacuole; subcellular distribution; body burden.

8.1 INTRODUCTION: DISTRIBUTION OF CHEMICALS IN ORGANISMS

Contaminants that have been taken up by an organism are either metabolized or stored (Figure 8.1). Because of this, the total body burden (including both metabolizable and stored chemical) can be much larger than the toxicologically relevant body burden. In order to be able to relate the two, one must be able to estimate the extent to which a chemical is stored in inert material and the extent to which a chemical is liberated from storage sites. Metabolism includes both the toxic effects (Chapter 11) and detoxification (Chapter 9). In a way, storing compounds as inert material can also constitute a mechanism of detoxification (see sections 9.2 and 9.3). The shift from inert to metabolizable compound may cause delayed toxicity

An Introduction to Aquatic Toxicology
http://dx.doi.org/10.1016/B978-0-12-411574-3.00008-6

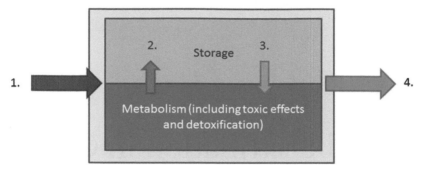

FIGURE 8.1   **A black-box model of the fate of a chemical in an organism.** (1) A chemical taken up by an organism can either be directly metabolized or (2) stored in tissues such as fat tissue and the skeleton. (3) Delayed toxicity can be caused when a chemical is liberated from the storage site. (4) Finally, the chemicals are excreted either unchanged or as biotransformed daughter compounds.

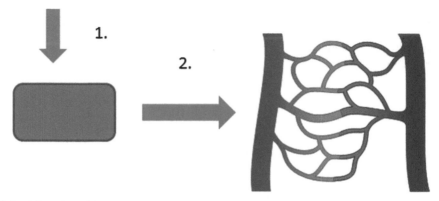

FIGURE 8.2   **Principles of delayed toxicity.** (1) A stressor (e.g. starvation, acidification) causes (2) the liberation of toxicants (lipid-soluble ones from fat tissue, and hydrophilic ones, e.g. from the exoskeleton) to the blood stream with consecutive toxic effects on sensitive tissues.

(Figure 8.2). This is the case when lipophilic compounds are first stored in fat tissue, and then when the animals starve, lipophilic toxicants (e.g. polychlorinated biphenyls (PCBs) and dioxins) are released to the bloodstream. This has been a major reason for toxicity in, for example, reproducing seals. In carrying mothers, a clear increase in energy consumption occurs. As a simultaneous increase in food consumption cannot occur, fat deposits are used up. The fatty acids, diglycerides, and other lipid-soluble compounds, including lipophilic toxicants, are liberated into the bloodstream, whereby they can cause toxic effects in their target tissues.

8.2 STORAGE SITES OF CHEMICALS IN ORGANISMS

The major storage sites of contaminants in prokaryotes and plants are vacuoles, and in animals, the skeleton (both the exo- and endoskeleton) and fat tissue/fat bodies. Lipid-soluble substances are stored in fat tissue whereas ionic compounds are often stored in

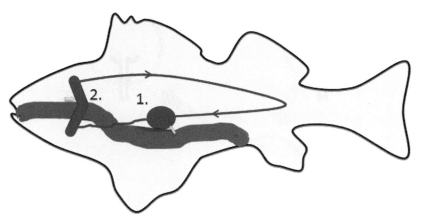

FIGURE 8.3 **A schematic account for why a chemical taken up in the gills has a different fate from that taken up in the intestine.** (1) A chemical taken up in the intestine (green arrow) goes first to the major detoxification center of the body, the liver, and effective first-pass metabolism takes place. (2) In the case of uptake in the gills (green arrow), such effective first-pass metabolism does not take place. Red arrowheads in the figure indicate the direction of blood flow.

either the exo- or endoskeleton. As the skeleton (e.g. shells and exoskeletons of shellfish and crustaceans) contains calcium carbonate, their dissolution as a result of acidification may result in liberation of ionic contaminants into the bloodstream and consecutive delayed toxicity (Figure 8.2).

The route of toxicant uptake in animals affects the metabolism, distribution, and storage of chemicals. For example, several chemicals show very different behavior if they are taken up from food in the intestine from if they are taken up from water in the gills. This is due to the fact that toxicants taken up in the intestine initially enter the bloodstream and pass to the major detoxifying center, the liver (in fish) or hepatopancreas (in many invertebrates). Because of this, the toxicity (and actually the amount of chemicals) decreases as a result of first-pass metabolism (Figure 8.3). Such first-pass metabolism does not occur to the same extent with toxicants taken up in the gills or the general body surface.

8.3 CELLULAR DISTRIBUTION OF CHEMICALS

The subcellular distribution of chemicals is often defined operationally. Differentiation between chemicals contained in cellular debris, organelles, soluble proteins (in the case of metals, metallothionein-like and other proteins), the heat-sensitive protein fraction, etc., can be made. The operational nature of the different fractions is shown by "organelles" being one entity as distribution sites. However, organelles include both protein-producing entities (rough endoplasmic reticulum) and those where the breakdown of proteins occurs (proteasomes). The cellular distribution of substances can involve quite complex regulation, as described in Figure 8.4 for iron.

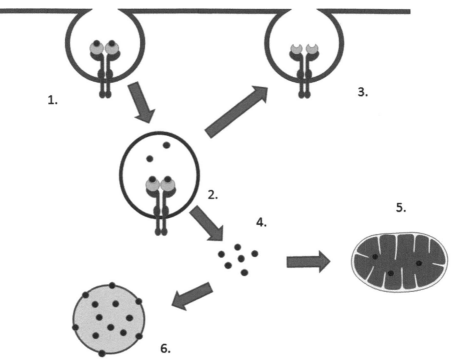

FIGURE 8.4 **An example of a complex behavior of a chemical in cells: the fate of iron the cell.** (1) Iron (gray-blue circles) is taken up mainly in the ferric state (Fe^{3+}) bound to transferrin (pale green), and captured and endocytosed with the transferrin receptor (red). (2) The endocytotic vesicle is acidified by proton-pump action (clathrin-coated vesicle), which results in iron liberation. (3) Iron-free transferrin receptor–transferrin complexes are recycled to the cellular membrane. (4) Iron crosses the endosomal membrane, probably in the ferrous (Fe^{2+}) state, and forms a labile cytoplasmic pool, from which it is either (5) transported to mitochondria for heme synthesis, or (6) is bound to ferritin. Ferritin-bound iron is the major iron store. In terms of the operational subcellular distribution, endosomes and mitochondria would be included in the organelle fraction, and ferritin-bound iron probably in metal-rich granules.

Relevant Literature and Cited References

Amiard-Triquet, C., Amiard, J.-C., Rainbow, P.S., 2012. Ecological Biomarkers: Indicators of Ecotoxicological Effects. CRC Press, Boca Raton, FL.

Blewett, T.A., Robertson, L.M., Maclatchy, D.L., Wood, C.M., 2013. Impact of environmental oxygen, exercise, salinity, and metabolic rate on the uptake and tissue-specific distribution of 17alpha-ethynylestradiol in the euryhaline teleost *Fundulus heteroclitus*. Aquat. Toxicol., 138–139. 43–51.

Camargo, J.A., 2003. Fluoride toxicity to aquatic organisms: A review. Chemosphere 50, 251–264.

Escher, B.I., Ashauer, R., Dyer, S., Hermens, J.L., Lee, J.H., Leslie, H.A., Mayer, P., Meador, J.P., Warne, M.S., 2011. Crucial role of mechanisms and modes of toxic action for understanding tissue residue toxicity and internal effect concentrations of organic chemicals. Integr. Environ. Assess. Manag. 7, 28–49.

Froehner, K., Backhaus, T., Grimme, L.H., 2000. Bioassays with *Vibrio fischeri* for the assessment of delayed toxicity. Chemosphere 40, 821–828.

Halappa, R., David, M., 2009. In vivo inhibition of acetylcholinesterase activity in functionally different tissues of the freshwater fish, *Cyprinus carpio*, under chlorpyrifos exposure. Drug Metabol. Drug Interact. 24, 123–136.

Nyman, A.-M., Schirmer, K., Ashauer, R., 2012. Toxicokinetic-toxicodynamic modelling of survival of *Gammarus pulex* in multiple pulse exposures to propiconazole: Model assumptions, calibration data requirements and predictive power. Ecotoxicology 21, 1828–1840.

Pickford, K.A., Thomas-Jones, R.E., Wheals, B., Tyler, C.R., Sumpter, J.P., 2003. Route of exposure affects the oestrogenic response of fish to 4-tert-nonylphenol. Aquat. Toxicol. 65, 267–279.

Detoxification

Abstract

Toxicants can be rendered less harmful by biotransformation, making non-harmful complexes, and then be stored in an inert state. Finally, the toxicant is excreted. Thus, detoxification can be considered to encompass events before excretion, although excretion as such is very closely linked with detoxification. The biotransformation pathway of organic compounds consists of phase 1, which generally increases the polarity of the compound, and phase 2, which usually involves conjugation to make excretable compounds. The most-studied system for biotransformation of organic compounds is the aryl hydrocarbon receptor (AhR) pathway, which is also called the xenobiotically induced gene-expression pathway. Dioxins are among the most-studied AhR ligands, so much so that the ligand-activated transcription factor (AhR) is occasionally called the dioxin receptor. The increase of polarity in phase 1 is the result of oxidation, reduction, or hydrolysis. The most important and largest phase 1 enzyme group are the cytochrome P450 enzymes. While generally inducing, high concentrations of xenobiotics can inhibit the transcription, translation, and activity of phase 1 enzymes. Also, although the aim of biotransformation is to decrease toxicity, occasionally this does not happen in phase 1. The reasons for this are that the compound cannot be handled by the enzymes, that the formed product is more toxic than the parent compound, or that toxic reactive oxygen species (ROS) are formed in the enzyme reaction. As a result of phase 2 conjugations, usually excretable organic ions are formed. In addition to increasing the solubility of compounds in aquatic media, conjugation decreases the probability of the compounds being reabsorbed to the body across the lipid membranes of, for example, cells of the intestinal lining. While enzymatic biotransformation reactions detoxify organic compounds, metals are detoxified via complexation with, in particular, metallothioneins in animals and phytochelatins in plants. Metallothioneins are short proteins, whereas phytochelatins are glutathione oligomers. Green plants, algae, fungi, and prokaryotes detoxify harmful compounds in vacuoles, where they remain inert. Detoxification by compartmentalization also occurs in animals, where compounds can be transferred to inert tissues.

Keywords: biotransformation; conjugation; phase 1; phase 2; aryl hydrocarbon receptor (AhR); dioxin receptor; cytochrome P450; mixed-function oxidases (MFOs); epoxide; glutathione-*S*-transferases (GSTs); sulfotransferases (SULTs); UDP-glucuronosyltransferase (UDPGT); metallothionein; phytochelatin.

9.1 BIOTRANSFORMATION OF ORGANIC COMPOUNDS

Virtually all organisms can transform organic toxicants to less toxic forms. Typically, detoxification enzyme activities are highest in the liver or equivalent organ in animals. Also, activities are higher in terrestrial than in aquatic organisms. This is because free diffusion of molecules out of the organism is possible for aquatic but not for terrestrial organisms. Biotransformation is normally a two-phase enzymatic process in which nonpolar lipophilic (organic) molecules are transformed to excretable polar and hydrophilic compounds. The best-known xenobiotically induced biotransformation pathway is the aryl hydrocarbon receptor (AhR) pathway for the biotransformation of compounds containing (aromatic) rings, which is present in virtually all organisms so far studied. This indicates that, although the AhR pathway has been studied mainly from the toxicological angle with focus on man-made chemicals, the ultimate biological reason for the presence of the pathway must be something else, although the toxicological angle has been prevalent for so long that the aryl hydrocarbon receptor is occasionally called the dioxin receptor. The scheme of AhR-dependent induction of gene transcription is discussed first. Then the principles of biotransformation, which can be divided into phase 1 and phase 2, and which are schematically illustrated in Figure 9.1, are

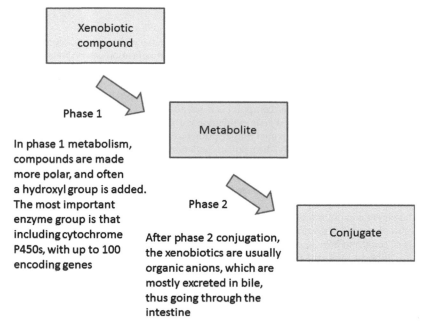

FIGURE 9.1   **The principles of detoxification of organic contaminants.**

given. Among aquatic animals, fish are the best-studied group, so the description below is mainly based on information from fish.

9.1.1 AhR-Dependent Signaling

The most important inducible biotransformation pathway of organic molecules is the AhR pathway. The induction of gene expression by AhR, also called xenobiotically induced gene expression, is given in Figure 9.2. AhR belongs to the bHLH-PAS family of transcription factors,

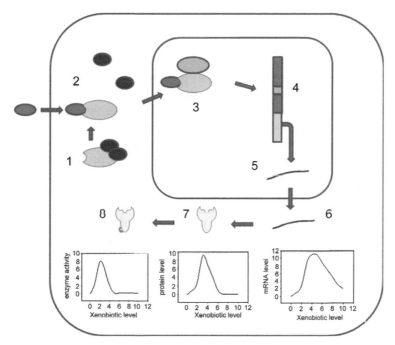

FIGURE 9.2   **The AhR pathway for induction of biotransformation-enzyme production by xenobiotics.** (1) In the absence of ligand, the receptor (green) binds a dimer of HSP90 (red), which acts to inhibit the proteolysis of the protein. In addition to HSP90s, several other proteins interact with AhR. (2) Upon ligand binding, the HSP90s dissociate from AhR, and the ARNT dimerization site is exposed. The ligands with highest affinity are planar molecules with multiple benzene rings, such as dioxins and several polychlorinated biphenyls (PCBs); of naturally occurring compounds, ligands include at least tryptophan derivatives, bilirubin, arachidonic acid metabolites, and several dietary carotenoids. (3) The liganded AhR translocates to the nucleus, where it forms a dimer with ARNT (gray). (4) The dimer binds to the XRE (xenobiotic-response element, depicted in orange) in the promoter region of induced genes. (5) This causes the transcriptional induction of xenobiotic-responsive genes, cytochrome P450 1 (*CYP1*) being the most commonly studied. Transcriptional induction of *CYP1* is often seen with xenobiotic exposure. (6) However, if the concentration of AhR ligand is so high that it has a toxic effect on the transcriptional machinery, mRNA (red line) accumulation starts to decrease with increasing concentration of the chemical. The mRNA is translated to CYP1 protein, the level of which can be measured with antibodies. (7) If the sensitivity of the translational machinery to the toxic action of the xenobiotic is different from that of transcription, the dependence of protein amount on toxicant concentration will be different from that of transcription. (8) Finally, active CYP enzyme handles the xenobiotic, and again the toxicity of the AhR ligand towards enzyme activity may be different from that towards translation or transcription, since the activation of the enzyme may involve posttranslational processing (e.g. phosphorylation, depicted as an orange dot).

which respond to environmental and developmental signals. Apart from AhR, the group contains hypoxia-inducible-factor α subunits (HIFαs), the regulators of biological rhythms (CLOCK and BMAL), factors involved in neurogenesis (SIM), repressors of the AhR pathway (AhRR), and the general dimerization partner aryl hydrocarbon receptor nuclear translocator (ARNT). Initially, an organic molecule binds to the transcription factor aryl hydrocarbon receptor. In the unliganded state, the AhR is connected to heat shock protein 90 (HSP90). Upon ligand binding, the AhR–chemical complex is phosphorylated and transported to the nucleus, where an AhR–ARNT dimer is formed. The dimer binds to the xenobiotic-response element in the promoter/enhancer site of the target gene in the DNA. The binding causes the induction of gene transcription. In particular, the cytochrome P450 family of enzymes are the gene products encoded by the AhR-induced genes.

9.1.2 Phase 1

Phase 1 biotransformation either adds polar groups to organic molecules or exposes those already contained in a xenobiotic. Consequently, three types of reaction can occur: oxidation, reduction, or hydrolysis. Phase 1 biotransformation enzymes reside mainly in the endoplasmic reticulum, and their major classes are listed in Table 9.1. The most important group of oxidizing enzymes are the cytochrome P450 enzymes (often called mixed-function oxidases (MFOs)), which are present in all organism groups, including prokaryotes. There can be more than 100 different P450 enzymes in an organism, with some variation in the number between organisms. Many P450 enzymes can be induced by xenobiotics via the AhR pathway (commonly CYP1 enzymes). Although induction by toxicants usually occurs, long-term exposure or adaptation to high contaminant load can cause refractoriness, i.e. induction of the xenobiotic gene-expression pathway does not occur. Xenobiotic effects can be both transcriptional and posttranscriptional (described further in Figure 9.2). In addition, the enzymes play a role in eicosanoid, cholesterol, and steroid-hormone metabolism, and neural development. The overall oxidizing reaction of CYP enzymes is:

$$RH + O_2 + NADPH + H^+ \rightarrow ROH + H_2O + NADP^+ \tag{9.1}$$

Different classes of cytochrome P450 enzymes may interact. Another group of phase 1 oxidizing enzymes are the flavin-containing monooxygenases (FMOs), which were definitively characterized in the 1970s. Although the enzymes have only been purified from mammals, it appears that their toxicologically most-relevant substrates in aquatic animals such as fish are tertiary amines. When the enzymes catalyze the oxidation of sulfur-containing substrates, the resulting compound is typically more toxic than the parent one. In general physiology, the enzymes may be involved in salinity responses (when trimethylamine (TMA) levels typically vary). In addition, monoamine oxidase (MOA), alcohol and aldehyde dehydrogenases (ADHs and ALDHs), peroxidases, and aldehyde oxidases may play a role in phase 1 oxidations. Although reduction can occur in phase 1 of biotransformation, its importance is considered to be much smaller than that of oxidation. The reductive function of an enzyme usually depends on the availability of oxygen. The reductase activity is often associated with the production of oxygen radicals. An important reductase in xenobiotic biotransformation is the cytochrome P450 reductase, catalyzing the addition of single electrons to substrates such as quinones and aromatic amines. Other reducing enzymes are DT diaphorase, which appears to be induced through the AhR pathway, and azo- and nitroreductases, which appear to be important in the biotransformation of aromatic nitrogen-containing compounds. Hydrolysis deals especially with epoxides (which are often produced from a parent xenobiotic by cytochrome-P450-dependent oxidation). The major enzyme group is the epoxide hydrolases (EHs), which catalyze the addition of water to epoxides. They are mainly present in the endoplasmic reticulum, but some enzyme isoforms are present in the cytosol. The EHs present in the endoplasmic reticulum handle, in particular, the epoxides produced by the cytochrome-P450-dependent oxidation. Other substrates of EHs include epoxides of steroid hormones and juvenile hormone (in insects). Another important group of hydrolytic enzymes are the esterases—carboxylesterases and A-esterases. The esterases occurring in phase 1 were originally divided to A-esterases, which are activated by organophosphates, and B-esterases, which are inhibited by them. Carboxylesterases belong to the B-esterases. They are often membrane bound, and

TABLE 9.1 Major Phase 1 Biotransformation Enzyme Groups

Enzyme Group	Cofactor(s)	Remarks
OXIDATION		
Cytochrome P450	O_2, NADPH, cytochrome b	Present in all organisms studied. Usually a very diverse group with a multitude of substrates, e.g. zebrafish has at least 81 known genes encoding cyp enzymes
Flavin-containing monooxygenase	O_2, NADPH	Functions especially in biotransformation of nitrogen-containing xenobiotics
Monoamine oxidase	H_2O	
Aldehyde oxidase	NAD^+	
Alcohol dehydrogenase		
Aldehyde dehydrogenase		
Cyclooxygenase	Arachidonic acid, O_2	
Peroxidase	Peroxides (lipid-OOH or H_2O_2)	Synonyms lipoxygenase and PGH synthetase
REDUCTION		
DT-diaphorase	NAD(P)H	
Cytochrome P450 reductase	NADPH	The importance in xenobiotic metabolism of aquatic organisms very poorly known. Extensively studied in mammals.
HYDROLYSIS		
Carboxylesterase	H_2O	
Epoxide hydrolase		

catalyze the transformation of lipophilic organic esters to corresponding carboxylic acids and alcohols. They are present in all the animal groups so far studied. The A-esterases are important especially in the detoxification of organophosphate compounds (e.g. organophosphate insecticides). The A-esterases attack, for example, the O=P bonds, if they are in uncharged groups (it appears that ionized groups are not attacked). There is wide variation in A-esterase activity among animals. Consistent with the effectiveness of organophosphate insecticides, several insects do not have any enzyme activity in the non-acclimated state. The enzymes that attack organophosphate insecticides have Ca^{2+} as a cofactor. Thus, their activity is markedly reduced by calcium chelators. Although the previous discussion mainly concerns animals, especially fish, some phase 1 biotransformation enzyme activity is also present in plants.

Although the purpose of phase 1 reactions is to reduce the toxicity of xenobiotics and to transform them to end products suitable for conjugation in phase 2, occasionally detoxification does not occur (as schematically described in Figure 9.3). The first reason for this is that the toxicant is not suitable for detoxification reactions. This may be the case for high-molecular-weight (> 800) toxicants, and for toxicants containing C-halogen bonds. The second reason is that the biotransformation-produced chemical is more toxic than the parent compound. For example, MFO-catalyzed reactions change some organophosphorus insecticides

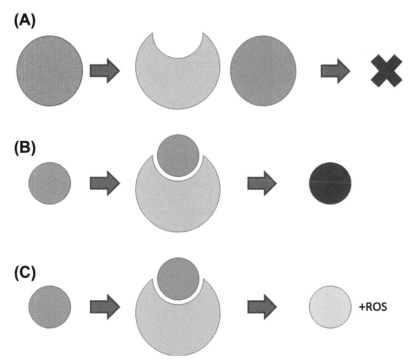

FIGURE 9.3 **A schematic representation of reasons why the biotransformation pathway may not lead to detoxification.** (A) The compound is too large for the phase 1 biotransformation enzyme to act on it, so the compound is not transformed. (B) The product of the phase 1 enzymatic reaction is more toxic (indicated by the more intense red color) than the parent compound. (C) Reactive oxygen species (ROS), which can exert toxic action, are produced in the enzymatic reaction.

to oxon-containing products with increased toxicity. Also, many epoxides produced by MFOs are more toxic than the parent compound. In fact, one of the most important functions of epoxide hydrolases is a protective function: hydroxylation renders the product less toxic. The third possibility is that reactive oxygen species (ROS) are produced in phase 1 biotransformation reactions. This is the case especially when reductases function in the presence of oxygen. Among ROS, the hydroxyl radical is the most reactive (and also the most short-lived, with a half-life of existence of about a nanosecond).

9.1.3 Phase 2

In phase 1, xenobiotics are rendered more polar. In phase 2, their polarity is further increased by conjugation with endogenous polar molecules, usually to form organic ions that can be excreted in aqueous solution. In addition to being important for excretion as such, the increase in hydrophilicity decreases the likelihood of the molecules being taken back up to the lipid membranes of, for example, cells lining the intestinal cavity. The major phase 2 conjugation enzymes are given in Table 9.2. There are three major conjugation pathways in phase 2. The conjugating enzymes are UDP-glucuronosyltransferases, sulfotransferases, and glutathione-S-transferases. In addition to these three types of conjugation systems, amino acid conjugation, acetylation, and methylation of compounds may take place in phase 2 biotransformation.

UDP-glucuronosyltransferases (UDPGTs) are membrane bound and occur especially in the endoplasmic reticulum (ER) membranes of the liver, with the active site facing the lumen of the ER. They are normally located in close proximity to phase 1 enzymes, which speeds up the whole biotransformation. The enzyme catalyzes the conjugation of uridine diphosphate glucuronic acid (UDPGA) with, in particular, hydroxyl groups (labile hydrogen) of both xenobiotic and endogenous lipophilic compounds. Sulfotransferases (SULTs) are largely cytosolic. They catalyze the transfer of sulfate from phosphoadenine phosphosulfate to the xenobiotic or endogenous substrate. Glutathione-S-transferases (GSTs) have many isoforms, and catalyze

TABLE 9.2   Phase 2 Conjugating Enzyme Groups

Enzyme Group	Cofactor(s)	Remarks
UDP-glucuronosyltransferase	UDPGA	
Sulfotransferase	3′-Phosphoadenosine-5′-phosphosulfate (PAPS)	
Glutathione-S-transferase	Glutathione	In addition to being a phase 2 conjugating enzyme group, GSTs are involved in the regulation of the redox balance (antioxidant defense)
Amino-acid-conjugating enzymes	Amino acids (especially laurine, glycine, and glutamine)	
Acetylating enzymes	Acetyl-coenzyme A	
Methylating enzymes	S-adenosyl methionine	

UDPGA, uridine diphosphate glucuronic acid.

the conjugation of reduced glutathione with electrophilic xenobiotics. As glutathione conjugation does not require hydroxyl groups, some organohalogen compounds and organophosphate insecticides can be processed. In addition to being phase 2 enzymes, some GST isoforms are integral components of antioxidant defense. The role of GSTs in antioxidant defense in aquatic animals may be more important than in terrestrial ones. Regardless of the conjugating enzyme of phase 2, the end products are anions and can be secreted in an aqueous medium.

9.2 DETOXIFICATION BY FORMING NON-HARMFUL COMPLEXES

The detoxification of metals often involves complex formation between the metal and sulfur-containing compounds. Probably the most-studied group of sulfur-containing metal-complexing compounds are metallothioneins. These are small proteins (61–68 amino acids), which have high numbers of cysteines (18–23); up to 30% of all amino acids in the molecules are cysteines, whereas they contain no aromatic amino acids or histidine. The sulfhydryl groups of cysteines react with metals. Virtually all organisms have genes encoding metallo-thioneins. There are altogether 15 families of metallothioneins; the proteins are divided to families mainly based on the organism type in which they are found (Table 9.3). The

TABLE 9.3   Metallothionein Proteins are Divided in 15 Families Based Mainly on the Organism in Which They are Found

Metallothionein Type	Organisms in Which Present, and Further Remarks
Family 1	Vertebrates
Family 2	Molluscs
Family 3	Crustaceans
Family 4	Echinoderms
Family 5	Diptera
Family 6	Nematodes
Family 7	Ciliates
Family 8	Fungi (type 1)
Family 9	Fungi (type 2)
Family 10	Fungi (type 3)
Family 11	Fungi (type 4)
Family 12	Fungi (type 5)
Family 13	Fungi (type 6)
Family 14	Prokaryotes
Family 15	Green plants (the family is nowadays divided into four groups)

transcription of genes is induced by most metals; the affinity of binding of several toxicologically much-studied metals to metallothioneins follows the following order: $Hg^{2+} > Cu^+ > Ag^+ > Bi^{3+} > Cd^{2+} > Pb^{2+} > Zn^{2+} > Co^{2+}$. Only a few protein structures in aquatic animals have been determined; most sequences given are ones predicted on the basis of nucleotide sequence. (Although the two are identical in most cases, the distinction should be remembered.)

The induction of metallothionein production is described in Figure 9.4. Notably, increased formation of metallothioneins depends largely on transcriptional induction. The binding of metal to metal transcription factor-1 (MTF-1) enables the factor to bind to the metal response elements (MREs) in the promoter regions of genes encoding metallothioneins; the binding increases transcription of the gene. Notably, although transcriptional induction via MRE occurs by Cd as well as by Zn, only Zn binds to MTF-1. There are several MREs in the promoter regions of genes. Differences in the number of MREs may explain some of the differences in the efficiency of induction. Also, only some MREs actually induce transcription, whereby the differences in structures of MREs between species and tissues can affect the degree of induction. Other reported factors influencing the accumulation of metallothioneins are translational efficiency and stability of the protein.

It appears that metals do not always cause metallothionein accumulation. In vertebrates, the induction is strong in the liver but much less effective in, for example, the gut and pancreas. In invertebrates, the induction is species-, tissue-, and time-specific. There is less available information on invertebrate metallothioneins than on those of vertebrates. From the fragmentary information available, it appears that there may be more metallothionein isoforms in invertebrates than in vertebrates. If the behavior of different isoforms towards metals were different, induction of one could be masked if its proportion were also affected by the treatment, as usually only the overall metallothionein level is determined.

Although the interest in metallothioneins has mostly been on metal detoxification by complexation, it is possible that the initial function of the proteins was to guarantee a steady and

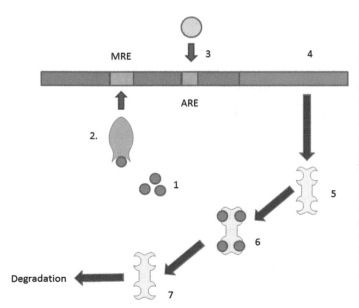

FIGURE 9.4 **The induction of metallothionein by metal (Zn) exposure and the metallothionein cycle.** (1) The Zn^{2+} ions (blue circles) accumulated in the cell (2) bind to metal transcription factor-1 (MTF-1), which attaches itself to the metal response element (MRE) in the promoter region of the metallothionein gene. (3) The induction of metallothionein gene transcription can also occur if electrophiles or, for example, H_2O_2 (green circle) bind to the antioxidant response element (ARE). (4) As a result of the induction, the metallothionein gene is transcribed and (5) apometallothionein is formed if the mRNA is translated. (6) When Zn^{2+} ions are present, they form complexes with metallothionein. (7) When the metallothionein protein loses the metal ions, it is degraded.

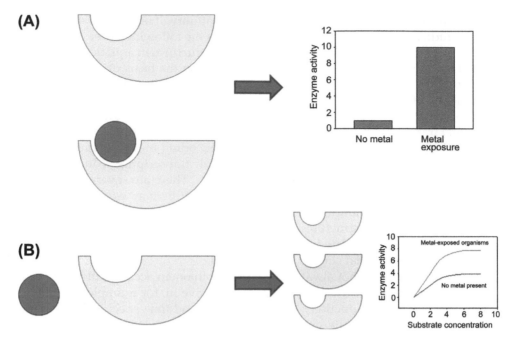

FIGURE 9.5 **Principles of the regulation of phytochelatin production as a result of metal exposure.** (A) The activity of phytochelatin synthase (phytochelatin production) is markedly increased by the presence (and probably binding to the enzyme) of metal. (B) The total activity of the enzyme (phytochelatin production) at a given metal concentration can be increased by the continuous presence of metal inducing the production of more enzyme protein. Such increased metal-induced gene expression has so far not been observed in animals producing phytochelatin.

adequate copper and zinc availability for use in enzyme synthesis; both metals are important constituents of the active groups of several enzymes. In this scheme, zinc and copper are taken up, bound to metallothionein, and stored until needed. When metals are needed in enzyme synthesis, the increased need can be fulfilled by liberation from metallothionein. The metallothionein level of actively dividing cells is much higher than nonproliferating ones.

In addition to being important in metal detoxification and metal homeostasis, metallothioneins can be involved in redox control. Their synthesis is induced by oxidative stress, and they scavenge oxygen radicals. Notably, when the metallothionein level was increased by cadmium exposure, the tolerance of the organism towards oxidative stress was improved, showing that the plentiful sulfhydryl groups of metallothionein are redox active.

In plants, phytochelatins are the major group of compounds that form metal complexes. Phytochelatins are oligomers of glutathione with 2–11 γ-glutamylcysteine moieties. Because the bond between glutamic acid and cysteine is not an amino bond, the molecule is not formed via ribosomal protein synthesis. Instead, metal exposure increases the activity of the enzyme synthesizing phytochelatin, phytochelatin synthase (γ-glutamylcysteine dipeptidyltranspeptidase (PCS)). Thus, while the enzyme is virtually inactive in the absence of metals, its activity is markedly increased when, for example, lead and cadmium are present. The principle of enzyme induction by cadmium is given in Figure 9.5. With metal exposure, the maximal activity of the enzyme can be increased by increased synthesis of the enzyme

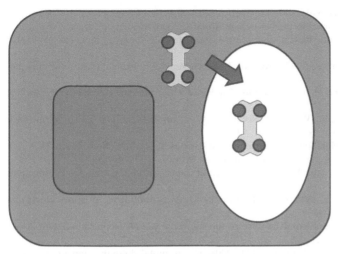

FIGURE 9.6 **An example of detoxification by compartmentalization.** The phytochelatin (light blue) produced in plants in response to metal (gray-blue circles) exposure is transported to a vacuole where it can be stored, and does not affect cellular functions.

protein. Although phytochelatins and their functions have best been described in prokaryotes, fungi, algae, and green plants, they have now also been found in several animal (invertebrate) groups. So far, metal induction of the protein activity only has been observed, and not an increase in the amount of the enzyme in animals.

9.3 DETOXIFICATION BY COMPARTMENTALIZATION

Green plants, algae, fungi, and prokaryotes detoxify harmful compounds by removing them to vacuoles in which they are removed from the general function of the organism. For example, it appears that phytochelatin is transported to vacuoles, whereby the complexed metals become inert (see Figure 9.6). Thus, high toxicant concentration can be measured without apparent ill effects. However, animals may also render potentially harmful substances nontoxic by compartmentalization. A common example of this is the storage of lipophilic compounds in fat tissue, as discussed in Chapter 8. Potentially harmful compounds can also be excreted to shells, the skeleton, fur, feathers, or scales, depending on the animal type. These excretion mechanisms are the same for both sexes, but by excreting toxicants in the egg mass, females can often decrease significantly more toxicants to gametes than can males, as the volume of egg mass is usually greater than that of sperm. This difference is greatest in mammals and birds.

Relevant Literature and Cited References

Ahearn, G.A., Mandal, P.K., Mandal, A., 2004. Mechanisms of heavy-metal sequestration and detoxification in crustaceans: A review. J. Comp. Physiol. B 174, 439–452.

Amiard, J.C., Amiard-Triquet, C., Barka, S., Pellerin, J., Rainbow, P.S., 2006. Metallothioneins in aquatic invertebrates: Their role in metal detoxification and their use as biomarkers. Aquat. Toxicol. 76, 160–202.

Andersson, T., Förlin, L., 1992. Regulation of the cytochrome P450 enzyme system in fish. Aquat. Toxicol. 24, 1–20.

Bainy, A.C., 2007. Nuclear receptors and susceptibility to chemical exposure in aquatic organisms. Environ. Int. 33, 571–575.

Blanchette, B., Feng, X., Singh, B.R., 2007. Marine glutathione-S-transferases. Mar. Biotechnol. NY. 9, 513–542.

Celander, M.C., 2011. Cocktail effects on biomarker responses in fish. Aquat. Toxicol. 105, 72–77.

Chambers, J.E., Yarbrough, J.D., 1976. Xenobiotic biotransformation systems in fishes. Comp. Biochem. Physiol. C 55, 77–84.

Claudel, T., Cretenet, G., Saumet, A., Gachon, F., 2007. Crosstalk between xenobiotics metabolism and circadian clock. FEBS. Lett. 581, 3626–3633.

Eljarrat, E., Barcelo, D. (Eds.), 2011. Brominated Flame Retardants. The Handbook of Environmental Chemistry, vol. 16. Springer, Berlin.

Gutierrez, A., Grunau, A., Paine, M., Munro, A.W., Wolf, C.R., Roberts, G.C., Scrutton, N.S., 2003. Electron transfer in human cytochrome P450 reductase. Biochem. Soc. Trans. 31, 497–501.

Ioannides, C. (Ed.), 1996. Cytochrome P450: Metabolic and Toxicological Aspects. CRC Press, Boca Raton, FL.

James, M.O., 1989. Cytochrome P450 monooxygenases in crustaceans. Xenobiotica 19, 1063–1076.

Malins, D.C., Ostrander, G.K. (Eds.), 1994. Aquatic Toxicology: Molecular, Biochemical and Cellular Perspectives. Lewis, Boca Raton, FL.

Miranda, C.L., Henderson, M.C., Buhler, D.R., 1998. Evaluation of chemicals as inhibitors of trout cytochrome P450s. Toxicol. Appl. Pharmacol. 148, 237–244.

Mommsen, T.P., Moon, T.W. (Eds.), 2005. Environmental Toxicology. Biochemistry and Molecular Biology of Fishes, vol. 6. Elsevier, Amsterdam, The Netherlands.

Perales-Vela, H.V., Pena-Castro, J.M., Canizares-Villanueva, R.O., 2006. Heavy metal detoxification in eukaryotic microalgae. Chemosphere 64, 1–10.

Sato, M., Bremner, I., 1993. Oxygen free radicals and metallothionein. Free Rad. Biol. Med. 14, 325–337.

Shariati, F., Shariati, S., 2011. Review on methods for determination of metallothioneins in aquatic organisms. Biol. Trace Elem. Res. 141, 340–366.

Varanasi, U. (Ed.), 1989. Metabolism of Polycyclic Aromatic Hydrocarbons in the Aquatic Environment. CRC Press, Boca Raton, FL.

Vergani, L., Grattarola, M., Borghi, C., Dondero, F., Viarengo, A., 2005. Fish and molluscan metallothioneins: A structural and functional comparison. FEBS. J. 272, 6014–6023.

Wang, L., Wu, J., Wang, W.N., Cai, D.X., Liu, Y., Wang, A.L., 2012. Glutathione peroxidase from the white shrimp *Litopenaeus vannamei*: Characterization and its regulation upon pH and Cd exposure. Ecotoxicology 21, 1585–1592.

Xu, Y., Morel, F.M., 2013. Cadmium in marine phytoplankton. Met. Ions Life Sci. 11, 509–528.

Zhou, H., Wu, H., Liao, C., Diao, X., Zhen, J., Chen, L., Xue, Q., 2010. Toxicology mechanism of the persistent organic pollutants (POPs) in fish through the AhR pathway. Toxicol. Mech. Methods 20, 279–286.

10

Excretion of Compounds from Organisms

Abstract

The excretion of toxicants is often called phase 3 of biotransformation/detoxification. Excretion includes both cellular excretion and excretion from the organism in the gills, kidney, and intestine mainly in bile. Cellular excretion can occur by diffusion for neutral lipophilic compounds, through different ion transporters for metal ions and small anions. For small ions, the efficiency of excretion depends on the relative affinities of the transporter for the toxic ion and its normal substrate. For most organic compounds, including organic anions and cations, the major way of excretion is via the ATP binding cassette transporters (ABC transporters). It appears that plants have the greatest number of different ABC transporter genes, and invertebrates may have more different transporters than vertebrates. ABC transporters use ATP to transport chemicals against their electrochemical gradient. Aquatic animals excrete hydrophilic compounds via the gills, small organic ions largely via the kidney, and most organic compounds as conjugates in bile. Bile formation depends on the secretion of bile acids to bile canaliculi, followed by osmotically obliged water. Xenobiotics are excreted to bile as conjugates largely via transporters of the ABCB type.

Keywords: bile acids; ABC transporters; urine; kidney; intestine; bladder; gills; multixenobiotic resistance (MXR); phase 3; ATP binding cassette; paracellular pathway; glomerular filtration; organic anion transport; phosphatidylcholine.

An Introduction to Aquatic Toxicology
http://dx.doi.org/10.1016/B978-0-12-411574-3.00010-4

10.1 INTRODUCTION

Excretion of toxicants from an organism (and its cells) is very closely connected to detoxification, as its ultimate purpose is to decrease the toxicant load. For this reason, excretion is occasionally termed phase 3 of the biotransformation/detoxification pathway. Also, when considering excretion, one needs to define whether excretion is efflux of toxicants from cells directly to the environment, to the intestinal lumen and the lumen of kidney tubules, or tubules of other excretory organs such as the Malpighian tube system within animals (i.e. formation of bile and its efflux to the intestinal lumen, or formation of primary urine and its flow to bladder), or efflux of toxicants to extracellular fluid (including the circulation and the aquatic environment). The possible excretion systems are depicted in Figure 10.1. This chapter discusses both cellular excretion mechanisms and excretion of contaminants from organisms. In the case of animals with intestine and excretory organs, the more complex excretory systems need to be considered. Again, since the most-studied aquatic animal group is fish, the discussion on excretion via different organs concentrates on this group. The distinction between intestinal excretion (feces) and urine/gill excretion depends mainly on the molecular weight of the secreted compound. With increasing molecular weight of a compound, it is increasingly secreted in bile to the intestinal lumen and finally in feces. If the molecular weight of a compound is less than 300, its likely secretion site is the urine or gills; for compounds with molecular weight of 300–600, the site depends on the other properties of the molecule; compounds with molecular weight above 600 are almost exclusively secreted in bile. The bile/urine ratio of a compound indicates what it preferential excretion pathway is.

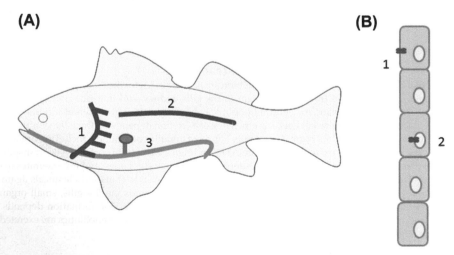

(A)   **(B)**

FIGURE 10.1 **The major excretory pathways in organisms.** (A) In aquatic animals (such as fish) excretion of chemicals can occur (1) via the gills directly to the aquatic environment, (2) via the kidney (or other excretory tissues) to urine, or (3) in bile (mainly formed in the liver or other similar tissues and excreted to the intestine). (B) In multicellular plants, compounds are excreted either (1) directly to the environment or (2) to vacuoles.

10.2 CELLULAR EXCRETION

The excretion of substances through cell membranes, which are the major barrier between the cytoplasm and any extracellular compartment (be it kidney tubule, bile duct, or aquatic environment) can occur by free diffusion, endocytosis, active or passive transporters of natural substrates, or active transporters belonging to the ATP binding cassette transporter (ABC transporter) superfamily. In the case of ions (e.g. metal cations and various anions), free diffusion is exceedingly slow, and consequently cellular excretion of these compounds only occurs in significant quantities if they are able to be transported by the different ion transporters present in the cell membranes. The likelihood of transport in this case depends on the relative affinities of the transporter for the normal substrate and the toxic one. The different possibilities of cellular excretion are schematically represented in Figure 10.2.

For organic compounds, including organic ions, the ABC transporters appear to be the major excretion route in all organisms, because they can actively transport substances against electrochemical gradients. Although the actual (rather than predicted) structure of ABC transport proteins has rarely been characterized in aquatic organisms, inhibitor studies, agonists, and cross-reactivity with antibodies made to mammalian transporters suggest that this is also the case in aquatic animals. In addition, more than 50 genes encoding proteins with predicted ABC transporter amino acid sequence are found in fish, including ones that are not found in terrestrial vertebrates. The presence of "new sequences" is probably due to genome-wide gene duplications that have occurred in teleost fish in comparison to mammals and other tetrapods. Cellular excretion of organic toxicants in aquatic invertebrates, such as crustaceans and molluscs, appears to involve the multixenobiotic resistance (MXR) mechanism (which is based on the ABC transporters). More than 60 genes encoding ABC transporters have been reported for *Daphnia pulex*. In comparison to animals, plants have many more genes encoding ABC transporters—about 120. A possible reason for the larger diversity of ABC proteins in plants than animals is that plants produce more toxic intermediates than animals, and these need to

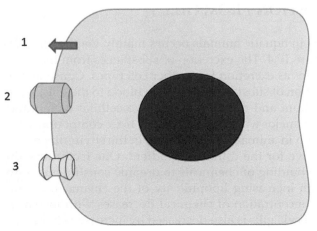

FIGURE 10.2 **The major cellular excretion systems.** (1) Excretion can occur via free diffusion for lipophilic compounds. The rate of diffusion is inversely proportional to the size of diffusing molecule. (2) Excretion can occur by specific ion transporters for metal ions and several small anions such as chloride, bicarbonate, carboxylates, etc., and (3) can occur by ABC transporters for most organic molecules.

be sequestered and transported to vacuoles where they are maintained in an inert state (see section 10.6 for a short section on excretion from multicellular plants). For example, it appears that metal-complexed phytochelatin transport from cytoplasm to vacuoles involves an ABC protein.

There are eight subfamilies (from ABCA to ABCH) of vertebrate ABC transporter genes, and all types are represented in *Daphnia*. The different functions of ATP binding cassette genes in animals are listed in Table 10.1 (with a major emphasis on mammals, as humans are by far the most-studied group). In addition to toxicant efflux, the ATP binding cassette proteins are involved in ion regulation, lipid metabolism, and immune functions in aquatic animals, and in, for example, nitrogen and phosphorus uptake in autotrophic organisms. Thus, the protein group, which is conserved from prokaryotes to mammals, has important functions in the normal physiology of organisms. Addressing these is beyond the scope of the present work, and only those involved in toxicant excretion are considered further.

The ABC transporters are either full transporters or half transporters. In the latter case, dimer formation is required for a transporter to be functional. The most common ABC transporter groups involved in xenobiotic excretion from cells are ABCC and ABCB. Of these, $ABCB_1$, the P-gp transporter, which is probably the most important animal xenobiotic transporter, accepts many different moderately (but not very) hydrophobic substrates, which often have planar structure and a basic nitrogen atom. The transport mechanism is as yet unknown, but may involve intercalating the hydrophobic substrate to the inner leaflet of the cell membrane, flipping of the molecule to the outer leaflet, and consecutive diffusion out of the cell. The tissue distribution of putative P-gp transporters suggests that food-borne toxicants may be an important reason for their evolution: reactivity towards antibodies is extremely high in the intestine. Also, gills are characterized with high reactivity. As it appears that flipping of the excreted substrate molecule from the inner to the outer leaflet of the cell membrane is an important component of the transport mechanism (see Figure 10.3 for the principles of the transport), direct consumption of ATP, which is in the form of (Mg) ATP, occurs.

10.3 EXCRETION FROM GILLS

Excretion of hydrophilic compounds in aquatic animals occurs mainly via the gills. The principles of excretion are given in Figure 10.4. The excretion of substances from the actual gill cells naturally has the same principles as excretion from other cell types. Consequently, the following discussion concerns excretion of substances carried in blood to the gills, which then must pass through the gill epithelium and minimally pass two cellular membranes. Diffusion out of the body via the gills is a major way of getting rid of toxic compounds. This naturally requires that their concentration in animals is higher than in the environment. The concentration in gill capillaries is decisive for the diffusion gradient. One factor affecting diffusible chemical concentration is the binding of chemicals to organic constituents in the blood. Since this binding increases with increasing lipophilicity of the chemical, the rate of diffusion as a function of the total concentration of chemical decreases with increasing lipophilicity. The excretion of compounds via gills is also decreased by increased charge (per

TABLE 10.1   ABC Transporter Gene Types and their Major Known Functions Based Mainly on Human Information

Gene (Capitals Human, Lower-case Other Vertebrates)	Known Protein (Mainly in Man)	Functions (Mainly in Mammals) and Other Remarks
ABCA1	ABC1	Cholesterol efflux
ABCA2	ABC2	
ABCA3	ABC3	Phosphatidylcholine efflux
ABCA4	ABCR	Retinyl efflux
ABCA5		
ABCA6		
ABCA7		
ABCA8		
ABCA9		
ABCA10		
ABCA11		
ABCA12		
ABCA13		
Abca14		
Abca15		
Abca16		
Abca17		
ABCB1	MDR, P-gp	Multidrug resistance protein; xenobiotic efflux
ABCB2	TAP1	Peptide transport
ABCB3	TAP2	Peptide transport
ABCB4		
ABCB5		
ABCB6	MTABC3	Iron transport
ABCB7	ABC7	Iron/sulfide transport
ABCB8		
ABCB9		
ABCB10	MTABC2	
ABCB11	SPGP	Bile salt transport
ABCC1	MRP1	Xenobiotic transport

(Continued)

TABLE 10.1　ABC Transporter Gene Types and their Major Known Functions Based Mainly on Human Information—cont'd

Gene (Capitals Human, Lower-case Other Vertebrates)	Known Protein (Mainly in Man)	Functions (Mainly in Mammals) and Other Remarks
ABCC2	MRP2	Organic anion efflux
ABCC3	MRP3	Xenobiotic transport
ABCC4	MRP4	Nucleoside transport
ABCC5	MRP5	Nucleoside transport
ABCC6	MRP6	
ABCC7	CFTR	Chloride channel; involved in osmoregulation in fish
ABCC8	SUR	
ABCC9	SUR2	
ABCC10	MRP7	
ABCC11		
ABCC12		
ABCD1	ALD	
ABCD2	ALDL1	
ABCD3	PMP70	
ABCD4	PMP69	
ABCE1	OABP	Elongation factor complex
ABCF1	ABC50	
ABCF2		
ABCF3		
ABCG1	White	Cholesterol transport
ABCG2	ABCP	Xenobiotic efflux
Abcg3		
ABCG4	White2	Cholesterol transport
ABCG5	White3	Sterol transport
ABCG8	Sterol transport	
Abch1		

Although the above table is based on human information, most ABC genes appear to have close homologs in all vertebrates and many invertebrates. The biggest differences between vertebrates are in the ABCA group, where only about half of the genes (of, at most, 17) are retained in all groups. In all other ABC groups, there are at least 70% homologous genes. The ABCB and ABCC groups of gene products are largely involved in xenobiotic transport to bile, and ABCG gene products in the transport of cholesterol and other sterols. The *Abch1* gene is unique to fish among vertebrates (however, the gene type is common in invertebrates), but its gene product function is not known.

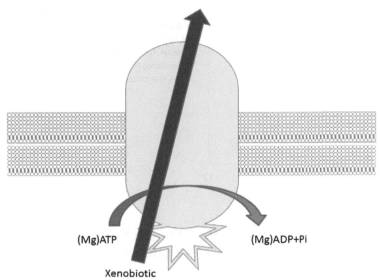

FIGURE 10.3   **The principles of ABC transporter function.** Pi signifies inorganic phosphate.

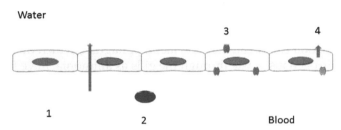

FIGURE 10.4   **The major excretion mechanisms of gills.** (1) Diffusion of lipophilic compounds can take place either through the cells or via the paracellular (between two neighboring cells) pathway. (2) In this case, an important limiting factor for excretion is the binding of molecules to the plasma or cellular constituents of blood. (3) Excretion of ionic compounds normally requires that the appropriate transporters are present in both blood-facing (basolateral) and water-facing (apical) membranes. (4) For some compounds, specific transporters are only needed at one side of the cell, if the electrochemical gradients and the properties of the molecule are such that free diffusion at one side of the membrane is possible.

unit volume of chemical). As illustrated in Figure 10.5, since decreased pH decreases the charge of weak acids, they are more efficiently excreted at low than at high pH values. If a lipophilic compound is ionized, it is usually not excreted in the gills, unless it is a substrate of the ABC transporters present in gill cells, but is excreted either in urine or bile. In addition to occurring through the cells, diffusion may occur paracellularly. Toxic metals and small anions can be excreted via diffusion, which is, however, very slow. Because of this, their major excretion occurs only if they are substrates of transporters (including channels that facilitate diffusion) located both in basolateral (blood-facing) and in apical (water-facing) cellular membranes.

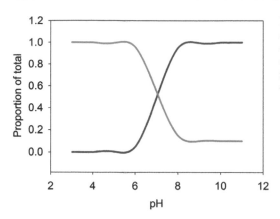

FIGURE 10.5 **The effect of pH on the proportion of a weak acid in ionic, base form (red line), and the effect of this on the proportional flux of the weak acid by diffusion (green line).** The rate of diffusion decreases with increasing pH until a constant value is reached, which indicates the rate of diffusion of the base form.

10.4 EXCRETION FROM THE KIDNEY AND OTHER EXCRETORY ORGANS

The excretion mechanisms in the kidney (and homologous organs of invertebrates) are depicted in Figure 10.6. The major functions of kidneys in terrestrial animals are to regulate the ion and water balance, and excrete nitrogenous waste. In aquatic animals, much of ion regulation and nitrogenous-waste handling is done in the gills, leaving regulation of the water balance as the primary kidney function. That the regulation of the water balance is the primary kidney function is indicated by the fact that hagfish, which do not regulate ion balance and excrete nitrogenous waste through the gills as ammonia, have functioning kidneys. The first step of urine formation is glomerular filtration. The blood pressure drives all compounds diffusively from capillaries to the glomerular space and consequently to the lumen of kidney tubules. The initial ultrafiltrate in the lumen of kidney tubules has virtually the same composition as blood plasma, with the exception that no proteins or other large molecules are present; all molecules with a radius less than 20 Å are filtered completely and those with a radius more than 42 Å are not filtered at all. The appearance of molecules in the ultrafiltrate also depends on their complexation with high-molecular-weight plasma constituents. If, for example, a compound is tightly associated with albumin, it will not appear in the liquid of the lumen of kidney tubules. The epithelial cells of the kidney tubule are able to secrete organic anions and cations to the lumen of the tubule with the help of appropriate ABC transporters. The concentration of any compound in the ultrafiltrate of the lumen increases with the reabsorption of water taking place in the loop of Henle and the collecting tubule—which occurs to a much greater extent in marine than in freshwater organisms. The same sites are also the primary sites for salt reabsorption, occurring to a greater extent in freshwater than in marine organisms. Finally, the formed primary urine enters the bladder, from which it is voided to the environment. Some water and salt reabsorption also occurs through the bladder epithelium. Notably, freshwater organisms produce copious amounts of dilute urine, as their major physiological challenge is to get rid of excess water entering the body, whereas marine organisms reabsorb most of the water, producing only a little, very concentrated urine, since their major challenge is preventing water loss. Because of this difference, if the amount of a contaminant excreted is given as a function of urine volume, markedly different results can be

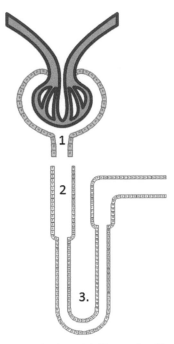

FIGURE 10.6 **Excretion mechanisms in the kidney.** (1) Glomerular filtration results in all small toxic molecules appearing in the ultrafiltrate in the lumen of the kidney tubule. (2) The endothelial cells of the proximal tubule are a site of xenobiotic excretion. (Mainly small organic anions are excreted in the lumen of the tubule.) (3) Water and salts are reabsorbed in the loop of Henle (and collecting tubule), whereby the concentration of xenobiotics in the primary urine increases.

observed between freshwater and marine animals. On the other hand, if the excreted amount is given as a function of time, the values can be similar.

10.5 EXCRETION IN BILE VIA THE INTESTINE

Bulky organic compounds are excreted in bile. Xenobiotic excretion in bile consists of the following steps:

1. Transport of the molecules to hepatocytes in the bloodstream. Most molecules in the blood have not undergone biotransformation, which is a major task of the hepatocytes.
2. Transport of the compounds into the hepatocytes. For neutral lipophilic compounds, free diffusion is possible, as the compounds are biotransformed in hepatocytes, whereby the diffusion gradient for the parent compound is maintained. Organic cations and anions are taken up by the various ABC transporters and by a separate class of organic anion transporters (OAtp).
3. Biotransformation (Chapter 9) of the molecules that have not been transformed earlier.
4. Formation of bile and its secretion to bile canalicules.
5. Excretion of bile to the small intestine.

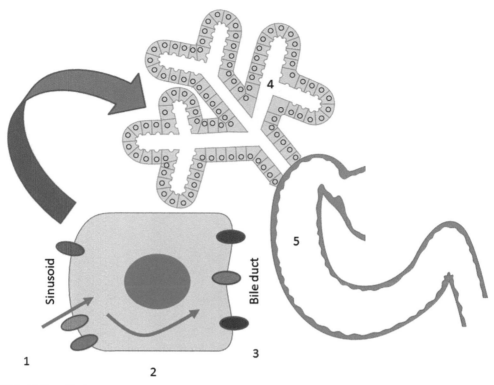

FIGURE 10.7 **Bile formation and secretion.** Bile is formed mainly in hepatocytes. (1) Hepatocytes take up bile acids using the Na-dependent taurocholate transporter, organic anion transporter, and ABC transporters (MXRs). These basolateral (sinusoidal) organic anion and ABC transporters are also used for xenobiotic uptake. Lipophilic compounds can also enter the cell by diffusion. (2) When in hepatocytes, xenobiotics undergo biotransformation. (3) Bile is formed mainly by the function of ABC transporters (MDR, MXRs, etc.). When bile acids are extruded from the apical (bile-duct-facing) membrane, water follows osmotically. Cholesterol and xenobiotics (mainly as conjugates) are then added to the bile. (4) Bile is excreted to bile canaliculi, blind-ended channels that form a tree-like structure ending in the largest bile duct, which empties into the small intestine (the early part of the intestine immediately after the stomach). The bile acids act as detergents, dissolving phosphatidylcholine from membranes to form a major component of the bile. (5) Bile enters the digestive tract in the early part of the intestine. Bile acids, as detergents, contribute to the digestion of lipids. The xenobiotic conjugates can be enzymatically digested to the lipophilic xeno-biotic and the hydrophilic conjugate components. If this happens, the lipophilic compound is reabsorbed and enters the cycle of bile formation again.

The sequence of events occurring in the secretion of organic molecules to bile is given in Figure 10.7.

10.5.1 Bile Formation

The primary bile contains mostly bile acids, cholesterol, and phosphatidylcholine, and, in addition, the xenobiotics that are excreted. The principle of cellular bile formation is shown in Figure 10.7. Bile acids are concentrated up to 1000-fold in hepatocytes as compared to the blood perfusing the cells. They are taken up into the cell mainly with the help of the sodium-dependent basolateral taurocholate transporter. The secretion of bile acid to bile canaliculi via the bile acid export pump is the primary step of bile formation, and the secretion at the apical membrane is the rate-limiting step. Bile acid secretion is followed by osmotically obliged water, and the water permeability of the cell is increased by the presence of aquaporins. The excreted xenobiotics are taken up either by diffusion followed by biotransformation, or by basolateral ABC transporters. The xenobiotic conjugates enter the bile by transport via apical ABC transporters, with P-gp being the most important. The bile acids act as a detergent, and dissolve phosphatidylcholine from the membrane of the canaliculus.

10.5.2 Secretion of Bile and the Further Fate of Conjugates Excreted in Bile

The secretion of bile, and the principles of its further fate in the intestine are shown in Figure 10.7. The formed bile enters the bile bladder in some species. In these animals, bile secretion is intermittent, and controlled largely by digestive signals. Bile acids are important in digestion of lipids—as shown in Figure 7.5, bile acids help micelle formation. In other species, the bile bladder does not exist, and bile secretion is thought to be continuous. Bile enters the intestinal lumen in the initial portion of the small intestine, immediately after the location of the pyloric caeca. After bile enters the intestinal lumen, the xenobiotic conjugates can be digested by appropriate enzymes in the small intestine. If this happens, a lipophilic toxicant can, again, be reabsorbed and cause toxic effects. In the case that the conjugates remain as conjugates, they form part of the intestinal excretion.

10.6 EXCRETION FROM MULTICELLULAR PLANTS (INCLUDING ALGAE)

Plants can excrete harmful chemicals by accumulating the chemicals to a part of the organism and then allowing this part to die and be removed. The rest of the plant can then continue living. Naturally, this applies only to multicellular green plants and algae, the latter of which are modular organisms with only little specialization in cell types. (However, one must bear in mind that, first, exposure via sediment and water is different; second, cell type differentiation to root, stem, and leaf cells has occurred; and, third, cells with different ages may respond differently.)

Relevant Literature and Cited References

Annilo, T., Chen, Z.Q., Shulenin, S., Costantino, J., Thomas, L., Lou, H., Stefanov, S., Dean, M., 2006. Evolution of the vertebrate ABC gene family: Analysis of gene birth and death. Genomics 88, 1–11.

Boyer, J.L., 2013. Bile formation and secretion. Compr. Physiol. 3, 1035–1078.

Burwen, S.J., Schmucker, D.L., Jones, A.L., 1992. Subcellular and molecular mechanisms of bile secretion. Int. Rev. Cytol. 135, 269–313.

Costa, J., Reis-Henriques, M.A., Wilson, J.M., Ferreira, M., 2013. P-glycoprotein and CYP1A protein expression patterns in Nile tilapia (*Oreochromis niloticus*) tissues after waterborne exposure to benzo(a)pyrene (BaP). Environ. Toxicol. Pharmacol. 36, 611–625.

Dean, M., Annilo, T., 2005. Evolution of the ATP-binding cassette (ABC) transporter superfamily in vertebrates. Annu. Rev. Genomics Hum. Genet. 6, 123–142.

Evans, D.H. (Ed.), 2009. Osmotic and Ionic Regulation: Cells and Animals. CRC Press, Boca Raton, FL.

Evans, D.H., Piermarini, P.M., Choe, K.P., 2005. The multifunctional fish gill: Dominant site of gas exchange, osmo-regulation, acid-base regulation, and excretion of nitrogenous waste. Physiol. Rev. 85, 97–177.

Kipp, H., Arias, I.M., 2002. Trafficking of canalicular ABC transporters in hepatocytes. Annu. Rev. Physiol. 64, 595–608.

Krasko, A., Kurelec, B., Batel, R., Muller, I.M., Muller, W.E., 2001. Potential multidrug resistance gene POHL: An ecologically relevant indicator in marine sponges. Environ. Toxicol. Chem. 20, 198–204.

Kullak-Ublick, G.A., Stieger, B., Hagenbuch, B., Meier, P.J., 2000. Hepatic transport of bile salts. Semin. Liver Dis. 20, 273–292.

Kurelec, B., 1992. The multixenobiotic resistance mechanism in aquatic organisms. Crit. Rev. Toxicol. 22, 23–43.

Luckenbach, T., Altenburger, R., Epel, D., 2008. Teasing apart activities of different types of ABC efflux pumps in bivalve gills using the concepts of independent action and concentration addition. Mar. Environ. Res. 66, 75–76.

Miller, D.S., 1987. Aquatic models for the study of renal transport function and pollutant toxicity. Environ. Health Perspect. 71, 59–68.

Miller, D.S., Pritchard, J.B., 1997. Dual pathways for organic anion secretion in renal proximal tubule. J. Exp. Zool. 279, 462–470.

Muller, M., Jansen, P.L., 1997. Molecular aspects of hepatobiliary transport. Am. J. Physiol. 272, G1285–G1303.

Navarro, A., Weissbach, S., Faria, M., Barata, C., Pina, B., Luckenbach, T., 2012. Abcb and Abcc transporter homologs are expressed and active in larvae and adults of zebra mussel and induced by chemical stress. Aquat. Toxicol. 122-123, 144–152.

Pritchard, J.B., Bend, J.R., 1991. Relative roles of metabolism and renal excretory mechanisms in xenobiotic elimination by fish. Environ. Health Perspect. 90, 85–92.

Pritchard, J.B., Miller, D.S., 1980. Teleost kidney in evaluation of xenobiotic toxicity and elimination. Fed. Proc. 39, 3207–3212.

Rea, P.A., 2007. Plant ATP-binding cassette transporters. Annu. Rev. Plant Biol. 58, 347–375.

Smital, T., Kurelec, B., 1998. The chemosensitizers of multixenobiotic resistance mechanism in aquatic invertebrates: A new class of pollutants. Mutat. Res. 399, 43–53.

Smital, T., Sauerborn, R., Hackenberger, B.K., 2003. Inducibility of the P-glycoprotein transport activity in the marine mussel *Mytilus galloprovincialis* and the freshwater mussel *Dreissena polymorpha*. Aquat. Toxicol. 65, 443–465.

Stieger, B., Meier, P.J., 1998. Bile acid and xenobiotic transporters in liver. Curr. Opin. Cell Biol. 10, 462–467.

Stieger, B., Meier, Y., Meier, P.J., 2007. The bile salt export pump. Pflugers Arch. 453, 611–620.

Sturm, A., Cunningham, P., Dean, M., 2009. The ABC transporter gene family of *Daphnia pulex*. BMC Genomics 10, 170.

Effects on Organisms

Abstract

This chapter outlines the major known mechanisms of toxicant effects at the level of the individual, focusing on animals. It discusses genotoxicity, reproductive disturbances, neural and developmental toxicity, teratogenesis and carcinogenesis, immunotoxicity, effects on energy metabolism and on membranes, and apoptosis. Oxidative stress as an important reason for toxicant actions is introduced. The chapter starts with an evaluation of genomic methods in aquatic toxicology. The possibilities and limitations of the microarray and quantitative polymerase chain reaction (PCR) methodologies, proteomics, metabolomics, RNA sequencing, and DNA methylation studies are detailed. Genotoxic effects range from point mutations to DNA strand breaks. A major reason for point mutations is that the influence of a toxicant on DNA overwhelms the DNA repair capacity. One reason for DNA damage is the formation of DNA adducts by many toxicants; another is toxicant-induced oxidative damage. With regard to reproductive disturbances, endocrine disruption caused by disturbances of reproductive hormone cycles as a major cause is discussed. Other important effects involve, for example, disturbances of sperm function and of development of the egg. Regarding neural effects, it is pointed out that insecticides are usually targeted to affect the nerve function of insects, and will influence the nerve function of other animals, too. An important effect on energy production is uncoupling—dissipation of the proton gradient required for aerobic energy production by mitochondria. Some toxicants affect the fluidity of cell membranes, with a consequent influence on the lateral movement of proteins in the membrane, and membrane permeability. Several toxicants cause apoptotic cell death. The chapter outlines the two (extrinsic and intrinsic) mechanisms of apoptosis. With regard to immunotoxicology, it is pointed out that the immune system as a whole should be understood, rather than concluding on the basis of a change in only one parameter that immune function is negatively affected by a toxicant. Similarly, the proximal mechanism that causes a toxic effect in embryos should be pursued, rather than lumping all observed disturbances into a common "developmental toxicity" entity, because the proximal reasons of effects appearing during development can be very different. Even teratogenesis (formation of abnormal structures) and carcinogenesis can have several causes. The chapter ends with a section on behavioral toxicity, as behavior is an integrated output of all effects on organismal, notably neural, function.

Keywords: genomics; transcriptomics; proteomics; metabolomics; gene ontology; microarray; quantitative polymerase chain reaction (PCR); RNA sequencing; microRNA; DNA methylation; mutation; comet assay; DNA adduct; DNA strand breaks; micronucleus; oxidative stress; reactive oxygen species (ROS); antioxidant; glutathione; endocrine disruption; vitellogenin; sperm motility; acetylcholinesterase; insecticide; neurotoxin; antidepressant; uncoupler; membrane fluidity; apoptosis; caspase; immunosuppression; phagocytosis; embryotoxicity; teratogenesis; carcinogenesis; feeding behavior.

11.1 'OMICS IN AQUATIC TOXICOLOGY—ECOTOXICOGENOMICS

Genomics can be defined as a discipline encompassing how the genes in the total genome of an organism function in the life-span of the organism. Based on this, ecotoxicogenomics can be defined as toxicant-induced changes in the genomes of organisms in an ecosystem. The different 'omics (transcriptomics, proteomics, and metabolomics) may be defined as belonging to genomics, and represent a set of methodologies that enable high-throughput data gathering from biological samples. They have resulted largely from two advances. First, equipment and methods have developed drastically during recent years. For example, the first bacterial genome sequence was released in 1995, that of a multicellular animal, *Caenorhabditis elegans*, in 1998, the human genome in 2001, and in 2013 there are about 10 sequenced fish genomes, close to 150 multicellular animal genomes, and about 100 plant genome sequences. The traditional sequencing machines are being replaced by next-generation sequencers (e.g. pyrosequencers), and the present level of methodological development enables rapid sequencing from a single DNA molecule. As a result, the plan of the Genome 10K Consortium—to have 10,000 genome sequences available—appears possible in the near future. Second, a whole new discipline, bioinformatics, has developed, with the aim of using the available computer power to handle the vast amount of data that is generated by the 'omics methods. One component of bioinformatics is the development and use of web-based databases that help in identifying the results of 'omics data, be it the nucleotide sequence of a gene, a peptide map of a protein, or a nuclear magnetic resonance (NMR) chemical shift in a compound. As with all information, the more data on a species are available, the more accurate and reliable any identification is. This section dwells mainly on methods (and things that need to be taken into account when applying them). This describes quite well the present situation of applying the latest 'omics methods to aquatic toxicology. One rushes out to use the latest methods without considering what conceptual addition they bring to understanding toxic actions. Thus, it appears that, instead of scientific innovation being the decisive factor, methods are the driving force. In fact, it is often said that 'omics approaches differ from earlier studies,

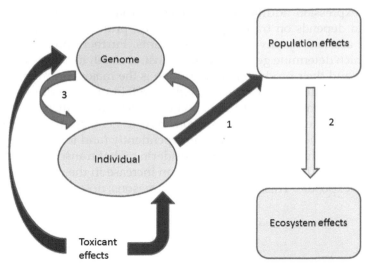

FIGURE 11.1 **The hierarchy of toxicant actions.** Toxicants always affect an individual (or its genome). (1) The effects on the function of individuals are seen across generations if the mortality or reproduction of a species is affected in such a way that the genomes of individuals are different after toxicant exposure from those not exposed. Population effects of toxicants are also observed only if toxicant exposure affects the reproduction or mortality of individuals directly or indirectly. (2) The effects on populations are reflected as ecosystem effects, since population changes will influence species interactions. (3) Effects on genomes of individuals are toxicologically important, if they are reflected in functional responses of organisms. Thus, one can say that a toxicant can have a biological effect only if individuals of some species respond to it, making the functional responses of individuals primary in the hierarchy of toxicant actions.

because they enable "discovery-based" work to be carried out without a priori expectations of what may happen. However, in the end it is the phenotype that responds to any toxicological insult. Figure 11.1 gives a scheme of the hierarchy of toxicant actions. It is often easier to work backwards from the phenotypic effects of toxicants to the possible genotypic influences than to try to guess the toxicological significance of changes in transcript expression. Taking this into account, a real need of bioinformatics is to develop tools that link genes, proteins, and metabolites. This problem is all the more challenging as the same end product (metabolite), can be produced via different pathways. The term "systems biology" has been coined as an expression for studies that try to evaluate how different pathways are involved in cellular (and organismal) function, with 'omics techniques and bioinformatics. Notably, this has always been the aim of physiology. However, the new methodological developments enable a far more versatile approach than was possible in traditional physiology.

11.1.1 Transcriptomics

11.1.1.1 Microarray and Quantitative PCR Studies

Presently, many if not most studies in aquatic toxicology use quantitative polymerase chain reaction (PCR) or microarray methodology. Normally, these methods give mRNA (or transcript) levels of the genes. This has led to the present-day use of "gene expression" as a synonym for "gene transcription" (and, surprisingly often, equating genomics to transcriptomics), although gene expression additionally includes translation, protein folding, and subunit aggregation, and depends on transcript, mRNA, and protein stability (see section 1.3). All of these can be affected by environmental toxicants. Furthermore, which phenomena are affected, and which determine gene expression most, are both toxicant- and gene-dependent. For some genes (and their products), transcription is the major controlling step; for others it is translation. In a small number of genes, either protein or transcript stability is the major control step. Consequently, even if full-genome correlations in man indicate that the most important regulatory step in the pathway from gene to gene product (= gene expression) is translation, every individual gene behaves independently (and toxicant-dependently).

The above has to be remembered when considering what transcriptional observations say about toxicant effects. The researcher observes an increase in the mRNA level of a gene as a result of toxicant exposure. This can be for several reasons, described in Figure 11.2. The first,

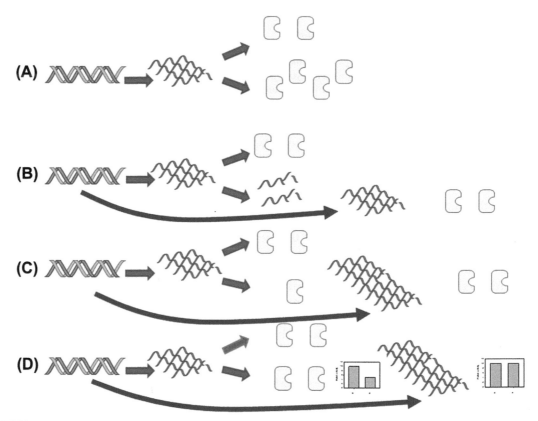

FIGURE 11.2 **A schematic representation of what an increased rate of transcription can mean.** (A) The toxicant increases the transcription of the gene with the result that more gene product is formed and its activity increased. (B) The toxicant reduces mRNA stability. Consequently, transcription must be increased to maintain protein production. (C) The toxicant causes decreased efficiency of translation or increased breakdown of the protein gene product. Consequently, transcription must be increased to maintain protein level. (D) The toxicant causes a decrease in the activity of a single protein, although the amount of protein remains constant. In this case, transcription must increase to maintain constant protein activity.

and the simplest one, is that the toxicant induces the gene in question such that the activity of the gene product is increased. Second, the toxicant decreases the formation of gene product, either by reducing translational efficiency or by reducing mRNA stability. In this case, an increase in mRNA level is required to maintain protein production at the required level. Third, the toxicant causes increased breakdown of the protein. Here, an increase in transcription is required to restore the protein level. Finally, the toxicant causes a decrease in the (unit) molecular activity of a single protein. Consequently, transcription must increase to maintain constant protein activity. Out of these possibilities, only the first would give a 1:1 correlation between transcriptional increase and gene product activity. This explains two major questions related to toxicant effects: first, one seldom observes a good concentration–response relationship in transcriptional data (see Chapter 12 for determinants of the concentration–response relationship); second, and it is hardly surprising, that most studies that have looked at transcriptional and functional correspondence report poor agreement between the two. The correspondence can be reasonably good when the system is in the steady state, but poor when acute stresses (non-steady-state conditions), which are usually characteristic of toxicant exposures, are applied. However, regardless of the reason, a change in gene transcription indicates that toxicant exposure has influenced the organism (see section 12.2). The change does not give the mechanism of toxicity, as from transcriptional change one cannot say what the reason for toxicity is. The toxicant may break specific bonds critical for the stability of the protein, or affect the properties of the active site in the protein.

The obtained transcript data are often treated as if the function of the gene product had been looked at. This possibility is made even easier as tools on the web (Gene Ontology, GO) allow this to be done without really considering what the basis is for assigning a gene to a function. Although Gene Ontology is species-neutral, it naturally supposes that similarity in the structure of a gene means similarity in the function of gene product. Thus, any functions of the gene product that differ from the most reported function are not easily picked up. Also, with an increasing phylogenetic distance from man and the commonly used laboratory rodents, mouse and rat, the proportion of genes (or actually the products of genes) with an unknown function increases. As an example, 1/3 of the genes in the *Daphnia pulex* genome could not be assigned a function. Notably, these genes are the ones that respond most to environmental perturbation, so their importance in aquatic toxicology is likely to be high.

As transcriptional changes indicate effects of toxicant exposure, they are useful for screening if exposure to a toxicant has occurred. This is especially so, since microarray methods enable the evaluation of all genes in an organism at one go. Thus, it is possible to find specific genes that respond specifically to the chemical used in the exposure, when it is taken into account that the response can be either immediate or develop more slowly (see Chapter 14). Important in the use of microarrays in aquatic toxicology is that the Minimum Information about a Microarray Experiment (MIAME) guidelines are followed and the data are deposited in a public database (http://www.ebi.ac.uk/microarray-as/ae/). Six components of microarray experiments must be reported in detail:

1. The design of the hybridization experiments as a whole.
2. The design of each array used and each element (spot, feature) on the array.
3. The samples used, extract preparation, and labeling.
4. The procedures and parameters of hybridizations.

5. The images used in measurements, and their quantification and specifications.
6. The types, values, and specifications of normalization controls.

   Notably, in early microarray work, often only one individual/pool of individuals was subjected to the array. This effectively abolished any biological variation occurring in the experiment. Biological variation is much larger in the commonly used laboratory fish, the zebrafish, than in rat and mouse. In natural populations of organisms, the variation is still larger; e.g. in zebrafish, for which data are available for both aquarium strains and wild populations, the difference is approximately fivefold. Consequently, even the common use of three replicates, while much better than one, gives a very poor picture of biological variation in toxicant exposures (see Figure 11.3 for the consequences of variability among individuals on obtained data). While biological variation of individuals in a population may be large and real, the different samples spotted to the microarray should all yield the same value. Thus, any differences in the values of technical replicates, with three again a common number of samples, represent measurement errors. The problem with microarrays is the cost, which in the past has been the main reason for the small number of biological replicates in the experiments.

   Making of microarray slides and the actual measurements are usually done in central facilities. The original articles described the microarray methodology in detail. One significant point is that while the methodology is especially versatile for species with fully sequenced genomes, of which there are only a few among the species commonly used in aquatic toxicology, developments allowing the use of the method in non-model organisms have been detailed. Also, several full-genome sequencing efforts are presently ongoing, with the decrease in the expense of full-genome sequencing. When full-genomic sequence is not available, microarray methodology may rely on first producing a complementary DNA (cDNA) library, from which expressed sequence tags (ESTs) are characterized by end-sequencing. When sequencing from the 5′ end, the directionally cloned cDNA often produces a nucleotide sequence of the protein-coding region of the gene. When the end sequencing is combined with subtraction, a procedure that enriches transcripts showing differences in representation between control and exposure, the number of sequences that must be cloned will be radically reduced. However, with normal subtraction, one is left with a large number of repeated sequences, making the amount of end cloning that needs to be done quite excessive. The number of redundant sequences is radically decreased when using the suppression subtraction hybridization (SSH) technique, which is an efficient means of producing cDNA libraries that contain mainly the transcripts that are affected by

FIGURE 11.3 **The differences between biological and technical variation.** (A) Biological variation. Only part of the variation in the responses to environmental change between organisms is covered. If the measured values do not represent the total variation in a population, they give a faulty picture of the effect. In the figure it can be noted that the three measured values do not give a full picture of the total responses to environmental change. (B) Technical variation. The same sample is measured several times, whereby all the values should be the same, and any differences between measurements represent measurement errors.

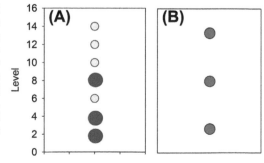

the exposure. The obtained cDNAs are mainly 3' ends, which are often noncoding. Capture of full-length cDNAs with biotinylated SSH fragments using streptavidin gives full-length cDNA clones for subsequent characterization, but retains the advantages of SSH. Alternatively, the availability of extensive EST data greatly assists in the identification of the SSH clones. The above discussion has concentrated on homologous microarrays, i.e. microarrays designed for the studied species. In aquatic toxicology, one often uses heterologous hybridization, utilizing arrays designed for another species. The success of this approach depends on the homology of the genes. Thus, one can expect that the closer the species are phylogenetically, the greater the overall success of heterologous microarrays. Commercially available zebrafish microarray slides have been successfully used to probe transcriptional changes occurring in temperature-stressed coral-reef fish.

The transcription pattern may be affected, for example, by the sex of the animals, the tissue, the time of day, and by the nutritional and developmental state of the organism. Consequently, defining experimental conditions accurately and with care is most important in order to evaluate the biological significance of the results. The effects of the above parameters on transcriptional activity are poorly known, so, at worst, microarray data in the absence of proper controls is quite inconclusive. Combining genomic observations with functional and structural ones is a major future challenge of genomics. Particularly in the case of invertebrates, many genes are not annotated and thus the functions of their products remain unknown. Furthermore, even if a gene is annotated, the problem remains that one does not know how translational efficiency varies between species, tissues, and developmental changes; all are points that are important when one is trying to relate transcriptional effects of toxicants to phenotypic and ecological toxicant responses.

Overall, there are significant positive reasons for the use of microarray technology in aquatic toxicology. First, the transcript expression of every gene in an organism can be theoretically determined. As a result, one can find transcriptional effects that represent completely new and unexpected affected pathways.

Further, large amounts of data are obtained. Consequently, new research questions/hypotheses about toxicant action can be generated and explored. Notably, both the above points can be regarded as starting points for new research. However, the methodology also has disadvantages, many of which are related to the facts that the methodology is quite expensive, and requires suitable skilled workers. Because of these factors, the methodology cannot be used in a regulatory context, although it would be potentially valuable, as one could expect that when the whole genome can be probed, different toxicants would give different transcript expression profiles. Mainly because of the costs, it is common to use only one or two time points. This is a major problem, since the transcription of genes as a response to toxicants may vary markedly with time, whereby snapshots at one or two time points may miss toxicant influence altogether. Also, microarrays in toxicology are often used without a clearly formulated research question; this can actually be understandable, as the overall "shotgun" approach gives the possibility of finding completely new genes that are associated with the toxicant. It is difficult to formulate a hypothesis on a topic that carries no previous knowledge. One should also remember on planning an experiment that if no new genes are expected/found to be associated with a toxicant as a result of a microarray study, no value is added to the much cheaper targeted transcript expression approach using qRT-PCR (quantitative real-time PCR; PCR arrays).

Because it is much cheaper than full-scale microarray, the most commonly used genomic method is qRT-PCR. Recently, PCR microarrays have been developed. In comparison to the traditional microarray methodology, which gives results for thousands of genes, PCR micro-arrays concentrate on defined pathways, usually with fewer than 50 genes. This gives two different departures for studying a toxicological problem. In PCR microarrays, one has as a premise that the toxicant is likely to affect the pathway considered. In the traditional microar-ray, which may target the whole genome of the organism, no a priori ideas about the affected pathways/functions are required. Often, quantitative PCR is used to verify microarray find-ings: in this case one takes some genes that are affected in the microarray and measures the mRNA-level changes observed with quantitative PCR. Usually, the two results are consid-ered to be in agreement if the direction and approximate magnitude is the same in both meth-ods. Critical for quantitative PCR is the normalization of the data. The most common method used is relating the mRNA level of the gene of interest to that of reference genes. This way of normalization gives relative quantification—the transcript amount of the gene of interest is related to the transcript amount of the reference gene. Observing absolute changes in the amount of the mRNA requires that the transcription of the reference gene remains constant (otherwise the results just show that the transcript amount of the gene of interest changes in relation to the reference gene; thus, depending on the reference gene, completely different results can be obtained, see Figure 11.4). In view of this, it is imperative that the constancy of transcription of the reference gene is evaluated in the experiment. This is particularly impor-tant for toxicological experiments as, in many cases, the transcription of even commonly used reference genes varies. It has been customary to take the reference genes from earlier (mainly

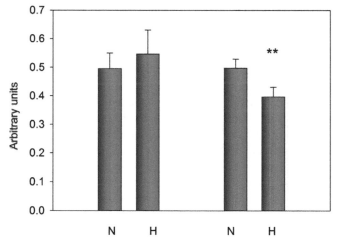

FIGURE 11.4 **The effect of reference gene transcription on the result of an apparent response of a gene to an exposure.** The amount of mRNA was determined using qRT-PCR with relative quantification using two different reference genes. Epaulette sharks were exposed to 2 hours' hypoxia (5% air saturation), and *HIF1* mRNA of cerebral samples determined: on the left, *HIF1* values were normalized to the transcript of myosin–phosphatase–rho interact-ing protein (*mrip* gene) and on the right, to the transcript of DNA J subfamily A2 (heat shock protein 40; *dnaja2* gene). While no change was observed for values on the left, *HIF1* mRNA appeared significantly reduced (** = P < 0.01, Mann–Whitney U-test) in hypoxia on the right (N = normoxic values, 10 animals; H = hypoxic values, 10 animals, bars indicate standard error of the mean). *Data from Nikinmaa and Rytkönen (2011).*

mammalian) work without testing for the constancy of transcription. Because the constancy of transcription of the reference genes cannot be guaranteed, reliable quantification can only be achieved by using two or several reference genes. Another possibility for normalization is external normalization. In this method, a known amount of mRNA is added to the sample at the onset of the qRT-PCR determination. Thus, the mRNA amount of the gene of interest can be related to the amount of added mRNA.

When transcript data are obtained, it must always be remembered that they do not give information about toxic actions unless they are linked to physiological/phenotypic effects. At present, it can be said that science has moved forward so that it is increasingly difficult to publish work that does not anchor transcriptional findings to observations of changes in protein (gene product) amounts and activities in exposures.

11.1.1.2 RNA Sequencing

The importance of RNA sequencing in aquatic toxicology stems from two facts. First, when the whole transcriptome is sequenced, one can get information about, for example, how different splice variants (of mRNAs) are affected by contamination. Second, noncoding RNAs (e.g. microRNAs) are important in posttranscriptional regulation of gene expression (they are involved in the regulation of translation of mRNA to proteins or mRNA stability, see section 1.3). If, for example, microRNA abundance is affected by a chemical, protein abundance may change even if the mRNA level is not affected. Consequently, the effect of the toxicant would not be seen in microarray or quantitative PCR measurement, but would be apparent in protein-level measurements or RNA sequencing. The choices of RNA sequencing range from sequencing the whole transcriptome (both coding and noncoding mRNAs) to sequencing only small noncoding RNAs. An increasing number of commercial providers offer RNA sequencing services with prices that are competitive. Thus, increasingly, the analyses can be done as a paid service, enabling the aquatic toxicologist to concentrate on the biological significance of the work. Also, guidelines as to what needs to be taken into account and reported in an RNA sequencing effort are available (Encyclopedia of DNA Elements (ENCODE) guidelines on RNA sequencing). Notably, in mammalian molecular biology, several microRNAs have been tied to specific conditions regulating translation. From the small number of studies available on organisms relevant for aquatic toxicology, it appears that microRNAs may be evolutionarily surprisingly well conserved, which suggests that mammalian findings can be related to even phylogenetically distant organisms. However, the significance of microRNA changes in aquatic toxicology is, as yet, little explored.

11.1.1.3 DNA Methylation Studies

The importance of methylation/demethylation in gene expression and epigenetic effects was introduced in Chapter 1. Most DNA methylation sites are situated on cytosines located 5′ to guanosines (CpG islands). Using bisulfite treatment (which changes non-methylated cytosines in CpG islands to thymine) and pyrosequencing (or PCR), the methylation state of a gene can be evaluated. The overall methylation state of DNA can also be evaluated using kits. However, it is difficult to link overall methylation to specific effects, because what is important is how individual genes are affected by toxic insults. Consequently, evaluating how the methylation of an individual gene is related to its transcription, and how methylation is affected by toxicants, are what give the most valuable new information. Although bisulfite treatment is the most common method used when DNA methylation is evaluated, commercial providers nowadays offer bisulfite-free methylation-determination kits. In this case, the determination relies on the differential cleavage of target sequences by two different restriction endonucleases, whose activities require either the presence or absence of methylated cytosines in their respective recognition sequences. Thereafter, real-time PCR is carried out. As real-time PCR quantifies the relative amount of DNA remaining after each enzyme digestion, the methylation status of individual genes (or set of genes) can easily be calculated.

11.1.2 Proteomics

The term "proteomics" has been coined to describe the characterization of proteins and their quantities. As with genomics, the field has become possible with the development of equipment, methods, and resources during the past 10–20 years. However, one needs to acknowledge that the number of fully characterized proteins is much smaller than the number of genes encoding them. What is given in the literature as protein sequence is often actually the amino acid sequence predicted on the basis of the nucleotide sequence of the encoding gene. Usually the two are the same, but one needs to remember the principal distinction.

Proteomic analysis typically starts with two-dimensional electrophoresis. In the first dimension, isoelectric focusing (IEF; often the ampholyte used has pH range 4–7) is carried out, whereby the proteins are separated mainly by their charge. The cylindrical IEF gel is loaded onto the second-dimension electrophoresis gel. The second dimension is normal electrophoresis, i.e. proteins are mainly separated on the basis of their molecular weight. Altogether, more than 1000 protein spots can be separated (see Figure 11.5 for the principle of two-dimensional

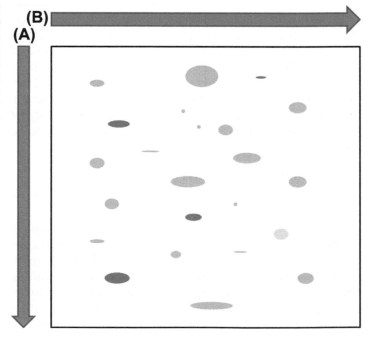

FIGURE 11.5 **The principles of two-dimensional electrophoresis for proteomic investigations.** (A) In the first dimension, isoelectric focusing is carried out, whereby proteins are separated mainly by their charge. (B) The second dimension is normal electrophoresis, i.e. proteins are mainly separated on the basis of their molecular weight.

electrophoresis). The characterization of proteins is done with mass spectrometry (usually matrix-assisted laser desorption/ionization with time of flight, MALDI-TOF), which analyzes peptide signatures present, and database searches finding identities of the peptide signatures of the protein spots. The characterization of a protein requires that the available databases have peptide sequences that have adequate similarity to the protein characterized. Consequently, protein identification is best for species (especially man, rat, and mouse) with many sequences deposited in databases, and the possibility for identification decreases with increasing phylogenetic distance from organisms with a large number of available protein sequences. The quantification of protein amount is done with normal densitometric analysis.

11.1.3 Metabolomics

Metabolomics is characterization of low-molecular-weight metabolites in biological samples, normally using either nuclear magnetic resonance (NMR) spectroscopy or mass spectrometry (MS). Consequently, similar to microarray technology and proteomics methods, the work is expensive and can only be done if the required equipment is available (although sample processing and determinations are relatively cheap when the equipment is available). When NMR and MS methods are compared, both have advantages, so neither can be advocated over the other. With regard to the number of metabolites that can be observed, Fourier transform ion cyclotron resonance mass spectrometry (FT-ICR MS), is the most sensitive. Furthermore, as the NMR method, which is most used in metabolomics, is nondestructive, the same samples can be used over and over again, or for other end points such as analysis of chemicals present. Depending on the compound, either ^{31}P or ^{1}H NMR is used. The NMR spectrum (i.e. the resonance peaks obtained) is unique for the set of compounds occurring, and the magnitude of the peaks shows the concentrations. The compounds produced in metabolism are more or less the same regardless of the type of organism. Since the metabolomics methods determine the results of metabolic activity, the obtained results are closer to physiological effects of toxicants than proteomics results (determining the amount but not the activity of the protein) or, especially, transcriptomics (microarrays or quantitative PCR; usually determining mRNA but not the protein amount).

11.2 GENOTOXICITY

Strictly speaking, genotoxicity refers to the effect of toxicants on the DNA structure of germ cells, spermatozoa and egg cells. In principle, genotoxicity means that any effects observed are transgenerational effects that occur via changes in DNA structure (in contrast to epigenetic effects, which can be defined as transgenerational effects occurring without changes in DNA structure). However, this principle has been broadened so that any effect on DNA is considered under this heading. The most often observed change is that point mutations occur (i.e. a single nucleotide in a gene is changed such that the coded amino acid in the gene product (protein) is changed). Also, in cells not exposed to any genotoxicant, DNA replication is associated with mistakes (nucleotide changes), which would form genes with point mutations. Because of this, the cells are equipped with effective DNA repair mechanisms, which correct most of the errors occurring. The presence of genotoxicants, however, increases the number of errors in the replication of DNA to such an extent that DNA repair is not adequate to remove all of them and, consequently, mutations can accumulate (see Figure 11.6 for a scheme of genotoxic effects). In addition to affecting the nucleotides coding for the amino acid sequences of the proteins, the changes may cause the introduction of inappropriate end codons, resulting in truncated gene products. The nucleotide changes may also affect the transcriptional start site, with the result that either extra amino acids are included in the protein or some amino acids normally encoded are not included in the gene product. However, since more than 95% of the DNA in the genome does not encode proteins, most of the nucleotide changes that genotoxicants cause occur in noncoding areas. These areas include binding sites for transcriptional regulators; consequently, genotoxicants can cause inappropriate activation of transcription or altered tissue distribution of gene products. Also, the areas that produce noncoding RNAs may be affected, causing changes in their structure. As, for example, microRNAs (a group of noncoding RNAs) regulate translation, genotoxicants can also influence the formation of gene products this way.

Some contaminants (particularly polyaromatic hydrocarbons (PAHs) such as benzo(a) pyrene, but also aldehydes) cause the formation of DNA adducts. The PAH molecule binds to a nucleotide in the DNA (guanosine is the most common base attacked, but adducts with other bases are also common). If the damaged DNA is not repaired (i.e. the base carrying the adduct is not replaced by an unmodified base), DNA cannot be replicated properly and mutation rate is increased. The increase in mutation rate depends on the type of DNA adduct. The occurrence of DNA adducts can be accurately tested, and thus can be used to indicate

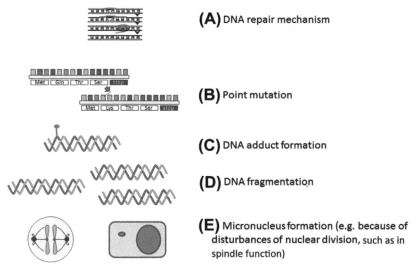

FIGURE 11.6 **The principles of genotoxic effects.** When the DNA repair mechanisms (A) are overwhelmed by toxicants, there is increased occurrence of mutation (B). This can occur as a result of increased formation of DNA adducts (C). Larger-scale genome disturbances may occur as a result of (D) DNA fragmentation or (E) micronucleus formation. Micronuclei are independent entities with maximally 1/3 of the size of the main nucleus.

exposure to mutagenic/carcinogenic compounds—the mutagenicity/carcinogenicity being actually directly related to DNA adduct formation.

Genotoxicants are more likely to affect actively transcribed, readily accessible DNA than the tightly coiled, inaccessible genome. Because of this, genotoxic effects usually take place in actively dividing cells. Since developing embryos have a large proportion of actively dividing cells, effects on DNA are more likely to be manifested during development than in adult life. By the same token, it is always more likely that the effect on DNA occurs in germ than in somatic cells, thus being a true transgenerational effect.

If effects on DNA occur in somatic cells, the likelihood of tumor production increases. Tumors are produced whenever the cells produced do not fit in the framework of the tissue where they are produced. This is usually the case when mutated cells are formed. Tumor formation is discussed in more detail in section 11.11. One talks about benign tumors when the likelihood of the production of metastases is small, and about cancerous growth when the cells forming the tumor continue to divide indefinitely and cells frequently escape the tumor and form new metastatic ones. The presence of tumors is common in aquatic animals stressed in any way. Thus, for example, freshwater fish (such as pike) living in a brackish water environment are much more likely to develop tumors than their conspecifics of the same age in freshwater.

In addition to point mutations that affect single amino acids in the gene products, or regulatory pathways as described above, effects on DNA may be larger. Toxicants may be clastogenic, i.e. induce breaks in DNA strands. For example, benzene and arsenic are known clastogenic agents. The genotoxic potential of contaminated water is often determined using the micronucleus test. This can be done with any cell type, as all studied shells have shown micronuclei either because of chromosomal damage (clastogenic effects) or aberrant nuclear spindle function. Aquatic vertebrates, invertebrates, and plants have all been studied. The equipment required is very simple: only a good microscope. The evaluation of micronuclei is usually done manually, but would also be possible automatically with suitable imaging software. Although any cell type can be studied, the test has been used especially on the circulating erythrocytes of fish. It should be remembered that the test measures genotoxicant exposure during the cell cycle (of cell production), which is normally markedly different from the life span of the cells. Because of the time taken for micronucleus production, the method is suitable essentially for chronic exposures, not acute ones. For example, with regard to fish erythrocytes, the micronuclei are developed during the division of the cells, and it takes at least a week before erythrocytes enter circulation. Consequently, one cannot see the effect of a toxicant on micronucleus formation in shorter exposures than this. If a toxicant affects the micronucleus proportion in shorter exposures, it is likely that the exposure affects the removal of cells with micronuclei in, for example, the spleen. It is known that the spleen removes micronucleated cells from circulation.

DNA is fragmented in apoptosis (see section 11.8 for a detailed discussion on apoptosis). Apoptosis is thus often evaluated using methods that determine DNA strand breaks (DNA laddering and the terminal deoxynucleotidyl transferase dUTP nick end labeling (TUNEL) assay). DNA strand breaks are commonly evaluated with single-cell gel electrophoresis (the comet assay). The comet assay is a technique for evaluating DNA damage (DNA strand breaks) in individual cells (see Figure 11.7 for the principle of the method). The overall DNA damage can be estimated from several measures of the proportions and relationships between comet head (intact DNA) and tail (fragmented DNA). While the different measures can be calculated manually, there are commercial

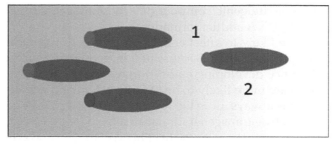

FIGURE 11.7   **The comet assay.** Single-cell electrophoresis is usually carried out in alkaline (pH > 13) conditions. Cells are encapsulated in low-melting-point agarose, on microscope slides. The encapsulated cells are lysed and electrophoresed. Intact DNA moves very little in electrophoresis, forming the head of the "comet" (blue, 1), whereas damaged DNA fragments move more, forming the tail (orange, 2). The overall DNA damage can be estimated from several measures of the proportions and relationships between comet head and tail.

sources for the entire comet assay. Direct microscopic examination of cells in metaphase or flow-cytometric analysis of cells can also be used to estimate DNA strand breaks. Sister chromatid exchange (SCE) refers to reciprocal exchange of DNA between sister chromatids. Although there is no actual DNA damage in chromatids that have undergone SCE, increased DNA strand break-age correlates with the appearance of SCE. The most pronounced effect on DNA is when whole or fragmented chromosomes fail to become incorporated to the nuclei of daughter cells, forming instead micronuclei. As discussed above, the reason for the production of micronuclei may be that the nuclear spindle does not work properly, with the result that not all chromosomal material will be located in daughter nuclei after cell division.

11.3 OXIDATIVE STRESS

An appropriate redox balance is required for proper cellular and organismic functions. The redox balance in the presence of oxygen is usually disturbed in the oxidizing direction. One can state that although oxygen is required for the life of all aerobes, it is highly toxic. Consequently, all aerobic organisms have developed effective mechanisms to combat oxida-tive stresses, i.e. disturbances of the redox balance towards the oxidizing direction. Oxidative stress can be defined as a disturbance in the pro-oxidant–antioxidant balance in favor of the former, leading to potential damage. The damage may be direct oxidative damage, but also indirect failure of any repair or replacement systems required to correct any damage. A hall-mark of oxidative stress is that the concentration of reactive oxygen species (ROS) increases. As their name indicates, these molecules readily react with available biomolecules, disturbing their function. There are marked variations in the stability and reactivity of different ROS (the different ROS are presented in Table 11.1). An increase may either be caused by increased ROS formation or by decreased efficiency of their removal, because of either decreased amounts of ROS scavengers (redox buffers such as glutathione and ascorbate) or decreased antioxidant enzyme activity. Notably, although ROS are usually considered as molecules associated with oxidative stress, they are increasingly shown to take part in normal cellular signaling. This increases the possibilities of ROS to be toxicologically important. Even at concentrations that do not cause measurable structural alterations, cellular signaling may be disturbed.

TABLE 11.1 The Major Reactive Oxygen Species

Compound	Chemical Equation	Remarks
Superoxide	$O_2^{\bullet-}$	The compound is produced e.g. in the electron transport chain of mitochondria (at atmospheric oxygen tension, 1–3% of oxygen reduced in mitochondria is converted to superoxide). The half-life of the molecule is of the microsecond order, and it is selective in its reactivity. It is the substrate of superoxide dismutase (SOD)
Hydroperoxyl radical	$HO_2^{\bullet}$	Formed in the dismutation of superoxide, and the reaction is often continued to form hydrogen peroxide and oxygen molecules
Hydroxyl radical	$OH^{\bullet}$	The radical is very reactive, but its short half-life of about 1 nanosecond restricts its movement to a few nanometers, whereby spatial heterogeneity in cells can be generated. The compound may take part in cellular signaling. It is formed in the Fenton or Haber–Weiss reactions, which involve ions of iron and copper
Peroxyl radical	$RO_2^{\bullet}$	Has a half-life of several milliseconds. Plays an important role in lipid peroxidation
Alkoxyl radical	$RO^{\bullet}$	Plays an important role in lipid peroxidation
Carbonate radical	$CO_3^{\bullet-}$	Formed when hydroxyl radicals react with carbonate or bicarbonate ions. It is a potent oxidant
Carbon dioxide radical	$CO_2^{\bullet-}$	
Singlet oxygen	$^{1}O_2^{\bullet}$	There are two types of singlet oxygen, of which the radical species $^{1}\Sigma g^{+}$ is rapidly converted to the non-radical $^{1}\Delta g$ state. Most references to singlet oxygen address the non-radical. The species is involved in photo-oxidation (photosensitization), which is inhibited by vitamin E
Hydrogen peroxide	H_2O_2	Causes cell senescence and apoptosis at concentrations above 10 μM (apoptosis is changed to necrotic cell death above 100 μM hydrogen peroxide). At low concentrations, the molecule may be involved in cellular signaling, as it e.g. promotes cell proliferation. Several enzymes, e.g. SOD, produce hydrogen peroxide. As such, the molecule, which is quite stable and very membrane permeant, is only weakly reactive, but its reactions with iron (and copper) produce the highly reactive hydroxyl radical
Peroxynitrite	$ONOO^{-}$	Attacks especially tyrosines of proteins, inactivating in this way several enzymes, e.g. SOD
Peroxynitrous acid	$ONOOH$	A strong oxidant and nitrant in aqueous solution, where it dissociates to peroxynitrite.
Nitrosoperoxycarbonate	$ONOOCO_2^{-}$	This compound is less reactive than peroxynitrite, most of which is converted to nitrosoperoxycarbonate in the presence of carbon dioxide

(Continued)

TABLE 11.1   The Major Reactive Oxygen Species—cont'd

Compound	Chemical Equation	Remarks
Hypochlorous acid	HOCl	Instead of being included in ROS, could also be called a reactive halogen-containing compound
Hypobromous acid	HOBr	Instead of being included in ROS, could also be called a reactive halogen-containing compound
Ozone	O_3	Ozone is produced when an oxygen molecule photodissociates to oxygen atoms, which further react with oxygen molecules. Causes inflammation and oxidizes lipids

Radical species are indicated in italics.

Antioxidants can be defined as any substance that significantly delays or prevents the oxidation of a substrate in an organism. The antioxidant defenses thus include:

1. Enzymes that directly remove free radicals.
2. Molecules that decrease the formation of radicals (this includes proteins that minimize the availability of pro-oxidants, such as transferrin binding ferrous ions and metallothionein binding copper).
3. Molecules that prevent oxidative damage to biomolecules (e.g. histones protecting DNA and chaperones such as heat shock protein 90 (HSP90) protecting proteins).
4. Small molecules that either quench pro-oxidants or are preferentially oxidized by them to leave the more complex biomolecules intact (redox buffers).

Small redox-sensitive molecules often contain sulfhydryl (-SH) groups, which are oxidized to intra- or intermolecular disulfide bridges. There are two types of nonenzymatic redox-sensitive molecules, endogenous and food-derived (see Table 11.2 for a list). The most important of all the small molecules is glutathione. It is present in most cells at millimolar concentrations. Since glutathione is present in all organisms, it is often measured to indicate the redox state of organisms. In this context, it is important to note that if glutathione is used to show the redox state of a cell, then the ratio between reduced and oxidized glutathione should be measured. The total glutathione content does not give information on the redox state. Figure 11.8 shows the glutathione cycle in its entirety. In addition to being important in redox regulation, glutathione is involved in conjugation of xenobiotics (phase 2 of detoxification, see section 9.1.3) in a reaction catalyzed by glutathione-S-transferase. While the several glutathione-S-transferase isoforms are, for the most part, not directly involved in combating oxidative stress, some can function in the prevention of lipid peroxidation. Major enzymes involved in antioxidant defense are outlined in Table 11.3. They can be broadly divided into enzymes involved in free radical or redox metabolism, and enzymes indirectly associated with redox changes. In addition to the enzymes listed, they include enzymes involved in the synthesis of the major small antioxidant molecules, such as glutathione (γ-glutamylcysteine synthetase (γ-GCS), glutathione synthetase) and ascorbate (the rate-limiting enzyme in synthesis; the enzyme lost in man and some other species is gulonolactone oxidase), enzymes involved in the formation of pro-oxidants, and enzymes regulating the equilibrium of redox couples (in addition to GSH/GSSG, the major ones are $NAD^+/NADH + H^+$ and $NADP^+/NADPH +$

TABLE 11.2 Nonenzymatic Components of Redox (Especially Antioxidant) Regulation

Compound	Remarks
Glutathione	The primary cellular redox buffer; discussed in more detail in section 11.3
Thioredoxin	Polypeptides with a molecular weight slightly over 10,000; present in all types of organisms. Thioredoxins exist in several forms, and their sulfhydryl groups react with several protein sulfhydryls, reducing them. They are enzymatically reduced back to –SH-containing molecules
Metallothioneins	Polypeptides with molecular weight of 6000–7000. In addition to metal sequestration, they are important in redox regulation (see section 9.2)
Caeruloplasmin	Can oxidize iron from the ferrous to the ferric state without ROS generation
Albumin	Provides accessible –SH groups to plasma components, and binds several redox-active compounds
Haptoglobin	Binds heme, whereby it cannot exert an oxidative effect
Bilirubin	The breakdown product of hemoglobin. Scavenges peroxyl and alkoxyl radicals, singlet oxygen, and peroxynitrite in vitro. In vivo influences are not known
Urate	Inhibits lipid peroxidation
Histidine-containing dipeptides	Since free histidine can function as a pro-oxidant, storing histidine needed for acid–base buffering in redox-inert dipeptides may be important for prevention of oxidative stress
Trehalose	A sugar accumulated in high concentrations by some organisms under stress; at high concentrations, can function as an antioxidant
Ascorbic acid	Although man and some other animals need to obtain this compound in food, most animals can synthetize it from available compounds
Glucose	Scavenges hydroxyl (and other) radicals. The use of the sugar in preservation is based on its antioxidant effect
Vitamin E (tocopherols)	α-tocopherol, the most important tocopherol, appears to be only a mediocre antioxidant. It may prevent lipid peroxidation, but many vitamin E effects may be independent from its antioxidant effects
Carotenoids (vitamin A)	Although they can, in principle, function as antioxidants in all organisms, a significant, definite role has been demonstrated in plants, where carotenoids prevent the formation of singlet oxygen in photosynthesis
Plant phenols	Are effective inhibitors of lipid peroxidation as the phenol groups effectively scavenge chain-breaking peroxyl radicals. Phenols can also scavenge several other ROS

Food-derived chemicals are given in italics.

H^+; a major enzyme affecting redox-couple balance is a rate-limiting enzyme of the pentose phosphate pathway, glucose-6-phosphate dehydrogenase (G6PDH)).

Figure 11.9 illustrates the different levels of oxidative stress. Oxidative damage in aquatic animals has often been assayed by measuring protein carbonylation and oxidation (as indicators of changes in protein structure), and lipid peroxidation using the TBARS

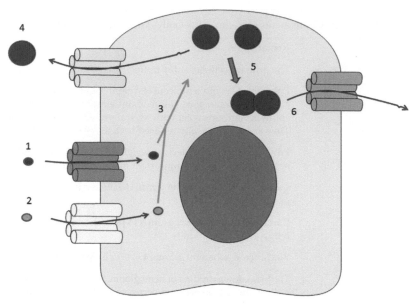

FIGURE 11.8 **The glutathione cycle.** (1, 2) The amino acids glutamate and cysteine are taken up into cells, either as such or as other amino acids subsequently metabolized in the cells into glutamate and cysteine. (3) Cellular glutathione is mainly formed enzymatically from glutamate and cysteine. (4) Some of the glutathione produced is excreted in the extracellular compartment or broken down. (5) In oxidizing conditions, e.g. oxidative stress, reduced glutathione (monomers) is converted to oxidized glutathione (dimers). Thus, the ratio between oxidized and reduced glutathione can be used as a measure of the redox state of the cells. (6) Oxidized glutathione dimers can be exported from the cells. Oxidative conditions may affect glutathione synthesis, breakdown, and efflux. The effects observed may depend on the cell type. Consequently, the total glutathione concentration of cells cannot be used to indicate the redox status of the cells. An increase, no change, and a decrease in total cellular glutathione concentration have all been reported upon oxidative stress.

TABLE 11.3   Antioxidant Enzymes

Enzyme	Remarks
Superoxide dismutase (SOD)	SODs are a large family of enzymes that dismutase superoxide to hydrogen peroxide and water. Typically they have metal ions in the active group, there being CuZnSODs, MnSODs, FeSODs, and NiSODs. Since the assay for SOD activity normally evaluates the overall activity, any changes in the distribution of enzyme isoforms remains undetected
Catalase	Catalases catalyze direct decomposition of hydrogen peroxide to water and molecular oxygen. It appears that catalases are most important when the hydrogen peroxide concentration is high
Glutathione peroxidase (GPx)	Glutathione peroxidases are a family of at least four isoforms. They remove hydrogen peroxide, reducing it to water while oxidizing glutathione. The reaction is $H_2O_2 + 2GSH \leftrightarrow GSSG + 2H_2O$. It appears that the importance of glutathione peroxidase in hydrogen peroxide removal increases with a decrease in the initial hydrogen peroxide concentration. Most glutathione peroxides have selenium in the active group

TABLE 11.3   Antioxidant Enzymes—cont'd

Enzyme	Remarks
Glutathione reductase (GR)	Oxidized glutathione (GSSG) needs to be reduced back to 2 GSH. This is done in a reaction catalyzed by glutathione reductase: $GSSG + NADPH + H^+ \leftrightarrow 2GSH + NADP^+$
Glucose-6-phosphate dehydrogenase (G6PDH)	The NADPH is provided by the pentose phosphate pathway, with G6PDH as the first enzyme. Another enzyme catalyzing NADPH generation is 6-phosphogluconate dehydrogenase
Glutathione-S-transferases (GSTs)	The enzymes are mainly involved in phase 2 of detoxification. However, some isoforms also appear to inhibit lipid peroxidation. The enzymes also affect redox balance indirectly, since the glutathione-conjugated xenobiotic is removed, whereby the redox buffer—glutathione—concentration is reduced
Thioredoxin reductase	Catalyzes the reduction of thioredoxins, so that they can be reoxidized and thus function to reduce target proteins
Peroxiredoxins	A family of peroxidases that can reduce organic peroxides and hydrogen peroxide
Other peroxidases	Reduce peroxides. Particularly rich variation in plants
Heme oxygenase	Removes heme, a strong pro-oxidant, and is involved in the production of the antioxidant bilirubin

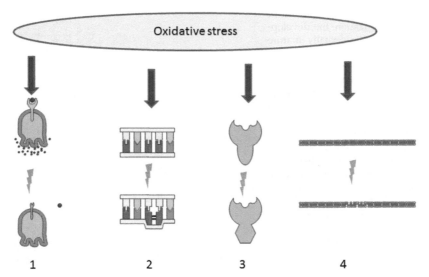

FIGURE 11.9   **Principles of oxidative stress effects.** (1) Before any structural changes occur, oxidative stress can affect signaling, since ROS appear to be involved in cellular signaling. (2) Oxidative stress can cause effects on DNA; for example, the formation of DNA adducts is increased. When damage to DNA overwhelms the repair capacity, an increased mutation rate is observed. (3) Oxidative stress can affect the three-dimensional structure of proteins, with the consequence that protein activity is altered. (4) Oxidative stress can influence lipids, causing, for example, lipid peroxidation. Such changes can cause alteration in the permeability of cell membranes.

(thiobarbituric-acid-reactive species) assay, and using the comet assay (as an indicator of damage to DNA structure). Oxidative damage can thus be observed in all major biomolecules. For example, the normal DNA structure can be disturbed with effects on DNA replication and consequent mutation rates (and tumor formation).

11.4 EFFECTS ON REPRODUCTION

Reproduction in aquatic organisms is, apart from a few exceptions, external. Thus, fertilization and the development of embryos take place in the ambient water. A flow chart of reproduction is given in Figure 11.10. Toxicants may have effects on:

1. The reproductive hormone cycles.
2. Gonad development, including the production and accumulation of the yolk protein vitellogenin.
3. Reproductive behaviors.
4. Fertilization.
5. The development of embryos.

The last can be considered as developmental toxicity, indicating the close connection between reproductive and developmental toxicity.

Toxicant effects on reproductive hormones have become almost synonymous with endocrine disruption, although in addition to reproductive hormones, stress hormones (cortisol/

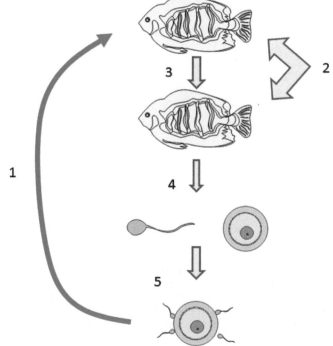

FIGURE 11.10 **Aquatic reproduction of animals.** (1) After fertilization, the development of an animal to the sexually mature stage occurs. (2) The reproductive cycle (i.e. the time taken between two consecutive reproductions) is mainly controlled by reproductive hormones. They are the targets of endocrine-disruptive chemicals (EDCs). The major environmental cues controlling reproductive hormone secretion are day length and temperature. (3) Gonad development depends on the hormonal reproductive cycle. Toxicants may accumulate in the developing semen and eggs. (4) Spawning. Its timing is controlled by reproductive hormones. (5) Fertilization. This depends on the properties of the sperm and eggs, which can both be influenced by environmental toxicants.

corticosterone, (nor)adrenaline), osmotic- and glucose-balance-regulating hormones (e.g. aldosterone, glucagon, insulin), feeding-regulating hormones (e.g. ghrelin, leptin, and orexin), and thyroid hormones (thyroxine and triiodothyronine) can be affected by different chemicals in vertebrates, and hormones affecting, for example, metamorphosis and building of the exoskeleton (juvenile hormone, molting hormone, molting-inhibiting hormone) can also be targeted in invertebrates. The endocrine disruptors affecting reproductive hormones can be, first, agonists of normal hormones (see Table 11.4 for a summary of endocrine disruption). Thus, they bind to the hormone receptors (e.g. estrogen and androgen receptors), and cause hormone-induced effects. An example of an estrogen agonist is diethylstilbestrol, which mimics the natural female hormone estradiol. Second, endocrine disruptors can be hormone antagonists. A true antagonist binds to the hormone receptor, blocking its function. For example, DDE (a metabolite of the insecticide DDT) can bind to the androgen receptor, and prevent normal testosterone function. Third, an endocrine disruptor can cause an increase in hormone synthesis. For example, aromatase activators, of which the herbicide atrazine may be the most important, are such endocrine disruptors. The aromatase enzyme is involved in testosterone–estrogen conversion (see Figure 11.11 for a schematic representation of steroid hormone metabolism). Fourth, enzyme synthesis can be decreased. Fifth, the breakdown of the enzyme can be either accelerated or slowed down. Sixth, endocrine disruptors can make an organism more or less sensitive to hormones later in life. It should be noted that enzyme effects may be involved in later embryonic development, which also brings developmental and reproductive toxicity together. Also, the effects of endocrine disruptors can be transgenerational if a chemical affects hormone function in such a way that alterations in the genome of germ cells occur (this includes changes in the methylation of DNA and histone modification). Many tests have been developed to assess the estrogenicity of aquatic samples. These include reporter assays with prokaryotes (see section 1.3.2) and vitellogenin determination (see below). In contrast, effects associated with masculinization (androgen-dependent effects) are much less studied and less well understood. One reason for this difference is that while there are almost universal female-specific biomarkers, male-specific ones are hard to find.

In addition to the compounds mentioned in Table 11.4, several organochlorine insecticides, phthalates, dioxins, and polychlorinated biphenyls (PCBs) are estrogenic, and organic tin compounds (tributyl tin and triphenyl tin) are androgenic. Reproductive hormones are important in the development of reproductive organs and gametes. Consequently, it is not clear if gamete development is directly affected by a toxicant or indirectly via a change in the hormone cycle. Significantly, however, estrogenic chemicals reduce milt production, which would be well in line with a disturbance in the hormonal cycle. The egg cells are much bigger than sperm, and contain yolk proteins that are used by the developing embryos for nutrition until exogenous feeding. The common precursor of all the proteins is vitellogenin, a glycolipoprotein that is cleaved to form yolk proteins. Vitellogenin is mainly produced in the liver of female fish and in fat bodies of invertebrates. Vitellogenin gene transcription is controlled by female hormones, estrogens. Also, the production of vitellogenin protein appears to be mainly transcriptionally regulated. For these reasons, the appearance of vitellogenin gene mRNA or produced protein in males is commonly used to indicate an exposure to estrogenic chemicals.

Aquatic animals usually lay their gametes in ambient water, where the sperm fertilizes the eggs. The milt is immobile until it encounters a hyper- (marine animals) or hypotonic

TABLE 11.4   Types of Endocrine Disruptor that Affect the Reproductive Hormone Cycles, and Their Principal Effects

Endocrine Disruptor	Effects
Estrogen agonists	Bind to the estrogen receptor and activate it. Compounds of this class include natural estrogens and artificial estrogens of contraceptive pills, phytoestrogens, cosmetics, and sunscreen compounds
Androgen agonists	Bind to and activate androgen receptors. The group includes natural androgens and some chemicals of cosmetics and sunscreens
Estrogen antagonists	Bind to the estrogen receptor, diminishing its activation by estrogens
Androgen antagonists	Bind to the androgen receptor, diminishing its activation by androgens. DDE functions in this way
Chemicals that affect other steps of the metabolism of natural hormones	
Chemicals that affect the production and breakdown of reproductive hormones	One important affected step is the function of aromatase, which is important in the catalysis of conversion of androgens to estrogens
Chemicals that affect the number of hormone receptors	

FIGURE 11.11 **Schematic representation of steroid hormone metabolism.** Apart from the cholesterol-side-chain cleavage enzyme and aldosterone synthase (synthesis of mineralocorticoids), which are mitochondrial, other conversion enzymes are located in the smooth endoplasmic reticulum. An important enzyme involved in androgen–estrogen conversion is aromatase.

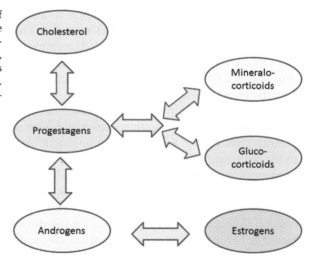

(freshwater animals) medium. In most cases, the salinity change initiates the motility. Ion movement through specific membrane channels is the proximal signal initiating the motility. Usually the motility is of relatively short duration, in some cases even less than a minute. The duration depends on the energy available: naturally, the longer the motility the more effective the fertilization. Thus, any toxicants that affect the motility of sperm will also affect the success of fertilization. As one of the factors affecting motility is energy production by the sperm, any toxicants affecting energy metabolism will affect the function of sperm and fertilization efficiency. This is probably behind the effect of tributyl tin (TBT) on sperm motility: the toxicant decreases mitochondrial energy production. Also, toxicants affecting the ion balance across membranes will affect sperm function. Metal effects on spermatozoan motility and function appear to be caused by the interaction of the metals with membrane transporters and consecutive changes in the signaling required for spermatozoan motility, including both cyclic AMP (cAMP)- and calcium-dependent signaling routes. Metal influx in spermatozoa can also affect the integrity of DNA when the metal ions cause oxidative stress. In contrast, estrogenic compounds do not affect the actual sperm function after milt production.

Fertilization involves the fusion of spermatozoan and egg cell membranes and the delivery of the haploid genome from the spermatozoan to the egg, where it combines with the haploid chromosomes of the mother. The decisive step is sperm activation, during which the thick egg membrane layers are penetrated. It appears that detergents that dissolve membranes are a group of toxicants affecting fertilization, by affecting the penetration of egg cells by spermatozoa.

11.5 NEUROTOXICITY

Neurotoxicity refers to situations in which an exposure to a chemical affects the function of the nervous system or peripheral nerves (or neuromuscular junctions). Most insecticides target synaptic transmission, especially at neuromuscular junctions. They are mainly excitotoxins. The normal chemical synapse functions in the following fashion: Neurotransmitters (the most important one at neuromuscular junctions being acetylcholine) are secreted from presynaptic neurons. The transmitters normally bind to the ion, especially sodium, channels of postsynaptic membranes. The binding causes the opening of the sodium channel and consecutive depolarization. This causes the postsynaptic cell to become excited. The excitation of the postsynaptic cell is finished when the neurotransmitter is enzymatically broken down, which results in the closure of the sodium channel. The most important neurotransmitter in neuromuscular junctions is acetylcholine, which is broken down in an acetylcholinesterase-catalyzed reaction. Chloride currents via GABAergic (GABA = γ-aminobutyric acid) channels oppose the excitation. Toxicants can thus either function at the ion-channel level or affect enzymatic neurotransmitter breakdown (see Figure 11.12 for the principles of synaptic transmission and the major targets for toxicant action). Among the insecticides that have been used, both types of excitotoxin functions have been designed. For example, pyrethroids and DDT target sodium channels preventing their proper closure, dieldrin and lindane facilitate excitation by decreasing GABAergic chloride flow, and organophosphate insecticides inhibit acetylcholine breakdown by inhibiting acetylcholinesterase activity. It is important to note here that, although they are called insecticides, the compounds will affect all animals with similar chemical synapses. The toxicity to a given animal type depends on the uptake and metabolism of the toxicant, and on the affinity of the site of action of the chemical in the particular animal. Every once in a while, insecticides are much more toxic to nontarget aquatic animals than to the target terrestrial insects.

Another group of neurotoxins that has recently become an issue in the aquatic environment is antidepressants, especially the selective serotonin uptake inhibitors (SSRIs) such as

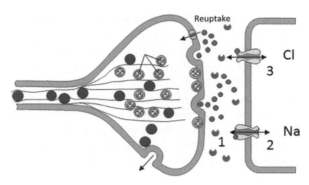

FIGURE 11.12 **Possible influences of neurotoxicants on the chemical synapse.** When the synapse is functioning normally, neurotransmitter (e.g. acetylcholine; red circles) is liberated in the synaptic cleft, and binds to the postsynaptic ion channels, affecting their activity. Normally, the transport of chloride and/or sodium is affected in such a way that the postsynaptic cell is depolarized, whereby it is electrically activated. The synaptic function is stopped when the neurotransmitter concentration decreases in the synaptic cleft either by reuptake or by enzymatic breakdown. As a result, the ion channel activity returns to a normal level. Many toxicants are excitotoxicants that prevent the return of the postsynaptic cell to the resting state. For example, (1) organophosphate insecticides inhibit acetylcholine breakdown by inhibiting acetylcholinesterase activity; (2) pyrethroids and DDT target sodium channels preventing their proper closure; (3) dieldrin and lindane facilitate excitation by decreasing GABAergic chloride flow.

fluoxetine. The mode of action of the SSRI compounds is considered to be the following: they increase the serotonin (5-HT) level of synapses by inhibiting serotonin transporters and thereby serotonin reuptake by cells, causing long-lasting elevation of the synaptic serotonin level. To have antidepressant effects, the plasma concentration in patients needs to be normally more than 50 µg/l. The neural mode of action has been accepted as the major effect. However, in aquatic invertebrates, reported effects are often in reproductive behavior or reproduction, e.g. induction of spawning, directly. Some reports suggest that these effects take place at presently observed environmental concentrations, which can be a few ng/l. Even with some bioaccumulation occurring, this is much below the concentration required for neural effects in mammals. If this is found to be a more common situation, then it is possible that a new effect pathway, independent of the traditional mammalian neural one, has been found.

Another group of neurotoxins is sedatives, with benzodiazepines as the most important group. These compounds are aromatic, and may cause their actions by influencing the fluidity of neural membranes. Thus, the majority of narcotic compounds affect cells in a fashion similar to ethanol, being membrane toxicants (see section 11.7).

11.6 EFFECTS ON ENERGY METABOLISM

Energy can be produced anaerobically in glycolysis (see Figure 11.13 for the different energy-production pathways), which takes place in the cytoplasm, or aerobically when glycolysis or the beta oxidation of fatty acids is followed by the Krebs cycle and oxidative phosphorylation. Aerobic energy production, which takes place in mitochondria, produces much more energy from a substrate molecule (often more than 10 times as much) than glycolysis. Largely because of this, toxicological studies on energy metabolism have

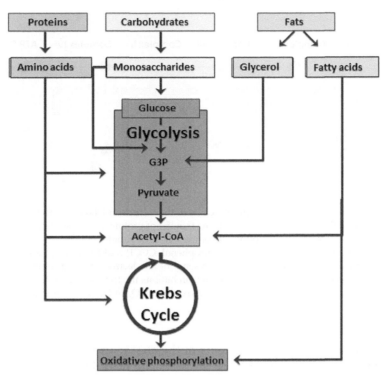

FIGURE 11.13 **Energy production by organisms.** Of the major substrates (proteins, carbohydrates, fats), fats can only be metabolized aerobically. Energy production involves anaerobic glycolysis, the Krebs cycle, and oxidative phosphorylation. The latter two occur in mitochondria, whereas glycolysis is cytoplasmic. G3P = Glyceraldehyde 3-phosphate.

concentrated on mitochondrial ATP production. In this context, one should note that as mitochondria have their own DNA, encoding some proteins, toxicants that influence the DNA may affect mitochondrial structure. Mitochondrial DNA appears, in most studies, to be more vulnerable to toxicants than nuclear DNA. As reasons for this, the close proximity to the electron transport chain, which produces ROS causing, amongst other effects, DNA damage; the lack of histones, rendering mitochondrial DNA more "naked" and accessible to toxicants than nuclear DNA; and the less effective DNA repair system than in the nucleus have been advocated. With regard to direct effects on energy production, the first group of toxicants are uncouplers (of oxidative phosphorylation). They dissociate the electron transport and phosphorylation reactions from each other (without affecting the proteins involved in either). A major class of uncouplers are hydrophobic weak acids, which can carry protons across mitochondrial membranes, and thus dissipate the pH gradient across the inner mitochondrial membrane, which is generated and actively maintained by the electron transport chain and required for the phosphorylation of ADP to ATP by the ATP synthase (proton pump). Examples of uncouplers are phenols, organic tin compounds, and other organometallic compounds. Another important group of toxicants affecting mitochondrial energy production are compounds that disturb the function of electron transport chain (respiratory chain) proteins and the proton pump.

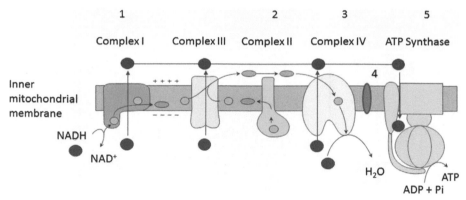

FIGURE 11.14   **The sites of action of major toxicants in the mitochondrial electron transport chain, involved in aerobic energy production.** (1) Rotenone inhibits the transfer of protons from complex I to ubiquinone. (2) Malonate and oxaloacetate are competitive inhibitors of complex II. (3) Cyanide and carbon monoxide inhibit complex IV (cytochrome oxidase). (4) Uncouplers (e.g. 2,4-dinitrophenol, CCCP, and tributyl tin) dissipate the proton gradient across the inner mitochondrial membrane. The presence of an electrochemical proton gradient is required for aerobic energy production by ATP synthase. (5) Oligomycin inhibits the function of ATP synthase. Red circles in the figure indicate protons, gray circles electrons, violet ovals coenzyme Q, and green ovals cytochrome c. Pi = inorganic phosphate.

These compounds affect different proteins of the chain, and Figure 11.14 gives examples of major toxicants and their sites of action. In addition to decreasing electron flow (and proton flow in the opposite direction), toxicants can affect the production of ROS (thereby being toxins that also target mitochondrial DNA). Glycolytic toxins mainly influence the enzymatic reactions. In particular, the reactions involving phosphorylation are targeted. For example, arsenic poisoning can be glycolytic: arsenate can replace phosphate in many reactions. Consequently, enzymes like hexokinase and glyceraldehyde-3,P-dehydrogenase can be affected.

11.7 MEMBRANE EFFECTS

Toxicants affecting membranes can act either on the lipid phase or on the membrane proteins. The actions on membrane proteins are usually protein-specific and, therefore, broad generalizations about these actions cannot be made. However, one important, common mode of action is inhibition of active ion transport across membranes. When active transport is inhibited, all the phenomena associated with actively maintained ion gradients, such as propagation of nerve impulses, secondarily active nutrient uptake, maintenance of intracellular pH, etc., will also be disturbed. The major active ion transport pathway is the sodium pump (Na^+, K^+ ATPase) that maintains sodium and potassium gradients across cell membranes. It is inhibited by cardiac glycosides (e.g. ouabain, digitoxin, digoxin). While these inhibitors are usually not environmentally relevant, the sodium pump is also inhibited by metals such as copper, mercury, cadmium, and lead in the micromolar range, and

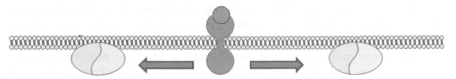

FIGURE 11.15 **The possible influence of toxicants affecting membrane fluidity.** Binding of a molecule to its receptor is followed by lateral movement of the receptor. Collisions of the receptor with the effector molecules activate the latter, and enable the responses. Toxicants that influence membrane fluidity influence the lateral movements of proteins, and affect the cellular responses generated.

by pesticides such as parathion also at micromolar range. Also, other insecticides such as organophosphorus compounds and malathion influence sodium pump activity, possibly by affecting the dephosphorylation of ATP. Metal effects may be associated with interactions of metal ions with cysteine sulfhydryl groups. Herbicides can also influence the function of the sodium pump. However, the effects are not likely to be ecologically important, as millimolar concentrations are required. The calcium pump, which is required to keep the intracellular calcium concentration in the low micromolar range, is affected by the same toxicants as the sodium pump. As many cellular functions use calcium signaling, which depends on a low intracellular calcium level, any disturbance decreasing calcium pump activity with the consequence that the intracellular calcium level increases, will affect calcium signaling.

In the lipid phase of the membrane, toxicants may affect the fluidity of membranes. Membrane fluidity determines how the proteins in the hydrophobic lipid sea are able to move laterally. Lateral movement of proteins is the basis of cell signaling and initial amplification of the signals. The fluidity of the membrane is one parameter that is similar in poikilothermic animals in the Arctic and in the tropics, when they are kept at their respective environmental temperatures. Since the fluidity of membranes increases with increasing temperature, this means that the membranes of Arctic organisms are more fluid than those of tropical ones, when fluidities are measured at a constant temperature (to increase the fluidity, cold-water fish have a much greater amount polyunsaturated fatty acids in their membrane than tropical ones). As, for example, cellular signaling can depend on lateral movement of membrane proteins, which, in turn, is influenced by membrane fluidity, chemicals influencing it will be membrane toxicants. A group of such compounds is ethanol and other alcohols, and another group is anesthetic drugs. Figure 11.15 gives a schematic summary of how and why toxicants affecting membrane fluidity can affect many aspects of cellular and organismal function.

Membranes (and cell walls) are the major barriers between the environment and organisms and between different compartments within the organisms. The thicker the barrier, the smaller the passive flux along the electrochemical gradient will be. Thus, all compounds affecting the thickness of membranes/cellular barriers will be toxicants via affecting gradients of ions and other cellular constituents between the environment or extracellular medium and cells. One particular group of such compounds is detergents, which dissolve lipid membranes. Lipid membrane barrier thickness of the gills, important both for respiration and ion regulation, is affected by many environmental toxicants, such as metals. Because of the importance of gills both in respiration and ionoregulation, these influences are discussed more fully in Chapter 16, which discusses the interactions between toxicant effects and natural environmental variability.

Another important property of cell membranes is their asymmetry. The double layer has different lipid species in the outer and inner leaflets: typically important outer-leaflet species are phosphatidylcholine and sphingomyelin, and those of the inner leaflet include phosphatidylethanolamine, phosphatidylserine, and phosphatidylinositol. The asymmetry of the lipid bilayer is actively maintained by flippase and floppase enzymes, and one signal causing removal of old and damaged cells is the appearance of inner-leaflet phospholipids on the outer surface of the cell, exposed to the environment. For example, the appearance of phosphatidylserine on the surface of the cells is associated with apoptosis and phagocytosis of the cells. Because membrane asymmetry is actively maintained, toxicants influencing energy production or consumption can also function as membrane toxicants.

11.8 APOPTOSIS AND NECROSIS

Apoptosis and necrosis are different types of cell death. Necrosis is defined as uncontrolled cell death, and is associated with cell swelling and inflammation. The hyperplasia typical for tissues damaged by toxicants is often the result of necrotic cell death. For example, the marked thickening of the gill epithelium associated with exposure to many toxicants is largely due to necrotic cell death. Apoptosis is programmed cell death. Many toxicants and other stresses such as hypoxic conditions may cause apoptotic cell death. The mechanism is not specific to any type of contamination. Initially, the cells become rounded and shrink, and detach from neighboring cells. Thereafter the membrane blebs, and small vesicles, called apoptotic bodies, are detached. The nucleus of an apoptotic cell becomes condensed and fragments. This nuclear fragmentation, DNA laddering, is used to indicate the presence of apoptotic cells. The TUNEL (terminal deoxynucleotidyl transferase dUTP nick end labeling) assay is a method often used to detect DNA fragmentation in apoptosis. The assay uses terminal deoxynucleotidyl transferase (TdT) to add secondarily labeled dUTPs to the ends of DNA fragments. The presence of the labeled fragments in cells can then be visualized microscopically. Inflammation does not occur in apoptosis. Two major apoptotic pathways are found, the so-called extrinsic and intrinsic pathways, and these are schematically represented in Figure 11.16. Two major differences are seen between the extrinsic and intrinsic pathways of apoptosis. Apoptosis via the extrinsic pathway is initiated by ligand binding to the death receptors. Typical death ligands are tumor necrosis factor (TNF) and Fas-ligand. Their binding to the respective receptors releases the active initiator caspase (caspase 8) from its interaction with FADD-protein, whereafter the effector caspases (cysteine aspartic acid specific proteases) carry out the major proteolytic activities. The intrinsic pathway is associated with the release of cytochrome C from mitochondria. The release is then associated with the activation of effector caspases and consecutive apoptosis. The main difference between necrosis and apoptosis is that whereas the former is associated with inflammation, this does not occur

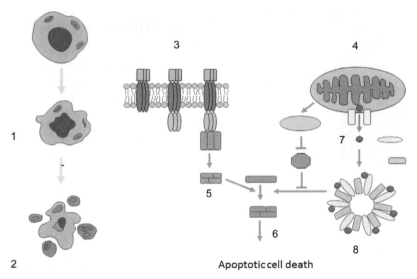

FIGURE 11.16 **Apoptosis.** (1) During apoptosis, cells usually shrink and detach from neighboring cells. (2) The membrane blebs and apoptotic bodies are formed. The nucleus condenses and the DNA fragments. Two major apoptotic pathways are found, the so-called extrinsic (3) and intrinsic (4) pathways. Apoptosis via the extrinsic pathway is initiated by extracellular ligand binding to the death receptors. This releases the active initiator caspase (5), which enables the effector caspases to be activated and carry out apoptotic breakdown of biomolecules (6). In the intrinsic pathway, mitochondria are damaged and cytochrome C is released from them (7). Together with some cytosolic proteins, cytochrome C forms apoptosomes (8), which function to activate effector caspases (6).

in apoptosis. In view of this, the apoptotic mechanisms may be as much geared towards preventing immune responses as towards executing cell death.

11.9 IMMUNOTOXICOLOGY

Immunotoxicology refers to the effects of toxicants on the immune system. To understand how toxicants can affect the immune system, it is necessary to have a general idea of immune system functions. If the immune system is defined as mechanisms preventing or decreasing the impact of foreign bodies on organismic function, then plants also have an immune system, although it is usually considered to be a specific property of animals. Plant defenses against foreign bodies are, first, aimed at preventing their entry into the organism, and, second, production of repellent compounds, especially terpenoids and phenols. Interestingly, the same compounds are used when plants respond to herbivore attacks. The basic principles of the animal immune system are given in Figure 11.17. Of the specific components, the adaptive (acquired) immune system (with lymphocytes) is only found in vertebrates. It is presently debated whether acquired immunity in any form is found in invertebrates. Regardless of the mechanism, the energetic costs of an effective immune mechanism must be balanced against the life length of the organism. Consequently, one can expect that the most complex invertebrate immune systems will be found in long-lived molluscs and crustaceans. Immunotoxicity in invertebrates has been studied particularly in molluscs,

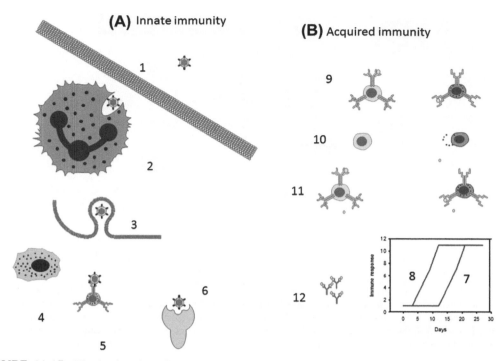

FIGURE 11.17 **The basic principles of the immune system in animals.** Immunity consists of (A) innate immunity and (B) acquired immunity, which is known certainly to exist only in vertebrates. Innate immunity is a property of the organism, and is not altered during its lifetime. Thus, the responses are similar in one-day-old and one-year-old animals. Innate immunity consists of (1) a barrier function: foreign particles are prevented from entering the body by barriers such as the skin. (2) If a particle enters the animal, it can be phagocytosed and broken down by phagocytosing cells. The invading particle can also be rendered inactive by encapsulation (3), or broken down by the action of natural killer cells (5) or proteins that break down foreign bodies (6). These proteins do not change in the lifetime of the organism. The inflammatory reaction, typically involving mast cells (4) and histamine release, is another aspect of innate immunity. In acquired immunity, the second exposure to a pathogen is followed by a faster immune response (8) than the first exposure (7). The immune response typically involves two types of lymphocytes, B- and T-lymphocytes (9) and humoral antibodies (12) secreted by these cells. The diversity of antibodies can be huge as a result of different combinations of subunits making up the antibody. In acquired immunity, a portion of cells, after the initial exposure to a pathogen, produce "memory cells" (10), which enable rapid response to a new exposure to a pathogen by helping a new population of specific pathogen-responsive cells (11) to be formed. Immunotoxicological research addresses the question of how toxicants affect any aspect of the above.

as they include several commercially important cultivated species. In the major group of aquatic vertebrates, fish, both innate and adaptive (acquired) immunity are present. Like other vertebrates, they have both B- and T-lymphocytes. The major difference from the mammalian immune system is the lack of defined lymph nodes or bone marrow. Many of the functions of the bone marrow are carried out by the head kidney (pronephros).

A general observation both in vertebrates and in invertebrates is that environmental pollution leads to immunosuppression. The prevalence of some bacterial and viral diseases has increased in polluted environments. In some cases, one has been able to induce increased incidence of disease by feeding contaminated food, e.g. PCB-contaminated herring to phocid seals. Another generalization that can be made is that low concentrations of several contaminants may cause activation of the immune system, while higher concentrations cause immunosuppression. Also, while tributyl tin is usually more toxic than dibutyl tin, it exerts less immunotoxic effect than dibutyl tin. A suggested explanation for this is that the immunotoxic effect may be calcium-mediated, and dibutyl tin is probably a better inhibitor of the calcium pump than tributyl tin, thus exerting a more pronounced effect. Apart from these generalizations, the components of the immune system are very varied. Consequently, one cannot state that a toxicant would affect the immune system as a whole; it may affect phagocytosis, encapsulation, ROS generation, antibody production, immune cell differentiation, etc., individually without affecting other aspects of immune function. On top of this, many aspects of immune function are affected by general stresses such as crowding, predator avoidance, temperature, oxygen availability, etc., making it nearly impossible to state that an observed immunological change is caused by contamination, unless a mechanistic link (cause–effect) between the contaminant exposure and the measured immunological parameter is established. (In this context it must be pointed out that a correlation between contaminant exposure and immunological response is not enough, as two correlated variables need not be associated to each other in any way, but the correlation between the two parameters may be the result of both depending on a third, undetermined variable). These caveats have been well taken into account by scientists doing immunotoxicological studies—invariably they emphasize that understanding the immunology of an animal as a whole is crucial for evaluating immunotoxicological effects and their causes. Presently, the most promising use of immunological parameters is as components of multibiomarker suites in establishing an ecotoxicological risk by a compound.

Chemicals that are known to affect the immune system include several metals, organometallic compounds (including dibutyl tin, described above), several pesticides, halogenated aromatic hydrocarbons such as dioxins and PCBs, and polyaromatic hydrocarbons such as benzo(a)pyrene. The most noticeable effect of metals is on ROS generation. The pesticides appear to affect both the innate phagocytosis and the acquired lymphocyte function. In vertebrates, both halogenated aromatic and polyaromatic hydrocarbons particularly affect lymphocyte proliferation. In invertebrates, halogenated aromatic hydrocarbons and PAHs appear to influence hematocyte stability and proliferation. While it is nearly impossible to estimate if the immunological effects of toxicants are observed at lower concentrations than other effects, their ecological significance may be great, as even a small disturbance in immune function may lead to increased disease susceptibility.

11.10 EFFECTS ON DEVELOPMENT

Developmental toxicity is any toxic effect that takes place during the development of an organism. The basic scheme of vertebrate development is illustrated in Figure 11.18. There are no toxicants that could be said to affect only development, rather the mode of action may be to affect the cell cycle, DNA duplication or repair, membrane behavior, etc. However, teratogenesis, discussed in section 11.11, is often taken as an aspect of developmental

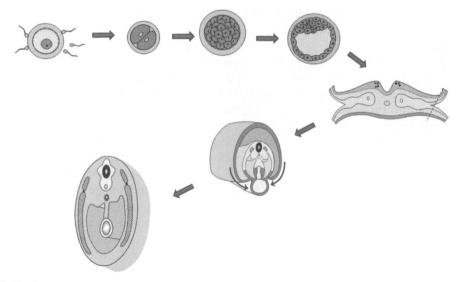

FIGURE 11.18 **Aspects of vertebrate development.** After fertilization, toxicants can affect development because active transcription of genes is much more common than in adults. Also, toxicants can influence cues (molecules) that affect tissue/organ development, causing, for example, teratogenesis.

toxicity, as developmental disturbances result in abnormal structures or functions. Another often-determined end point, which is solely of developmental origin and is often used in aquatic toxicology, is fluctuating asymmetry. This refers to asymmetrical properties of individuals, e.g. shells of molluscs. Their occurrence is taken to increase if an organism is developing in a contaminated environment. Quite often, embryos are considered to be more sensitive to toxicants than adults. Consequently, embryotoxicity is often evaluated and partial-life-cycle tests, which leave the time-consuming adult stage unexplored, are used when the toxicity of a compound is determined. The possible reasons why embryos should be more sensitive to toxicants than adults include the following. First, the rate of their cell division (and cell renewal) is much greater than that of adults. Thus, as many toxicants affect actively dividing cells and the proportion of them in developing organisms is much higher than in adults, the probability of a toxicant effect is also higher than in the adult. Since, furthermore, the structure of tissues is being formed, any disturbance will have a greater effect on the structure of a tissue than a disturbance of cellular function in an adult tissue. Second, many more genes are actively transcribed than in adults. Consequently, much more DNA is readily accessible to toxicants, which increases the possibility of genotoxic effects.

11.11 TERATOGENESIS AND CARCINOGENESIS

Many environmental pollutants can cause either the development of abnormal structures (teratogenesis; the occurrence of abnormal structures may lead to their abnormal function, and ultimately death) or the development of tumors. Often, both are caused by genetic effects—in the case of teratogenesis, the genome of the developing embryo has usually been affected already by the time that germ cells producing the embryo have been formed; in the case of tumor formation, the genotoxic effect has often taken place in the somatic cells of the adult. Consequently, one can classify teratogenesis to be within developmental toxicity. Any susceptibility to teratogenic agents depends on the original genotype and the type of toxicant exposure. The initiation of teratogenesis occurs at critical developmental periods. These "windows of susceptibility" occur during the times that the tissues where abnormal structures are found are formed. The teratogenic effects increase in frequency and severity with increasing amount of teratogenic agent. There are both naturally occurring and man-made teratogens. In some cases, a certain amount of a compound is needed for normal development, but an excess amount leads to teratogenesis; the compound functions to form structures at inappropriate places or times. Some teratogenic agents may also impair normal development by affecting the synthesis or function of molecules required for it. Teratogenesis is often tested using FETAX (frog embryo teratogenesis assay Xenopus), for which there are commercially available protocols.

Carcinogenesis refers to the formation of tumors (tumorigenesis) that are malignant. Increased cancer occurrence has been associated with aquatic contamination both in invertebrates (especially in molluscs) and in fish. The most significant association with contaminants is the occurrence of tumors in animals living in areas polluted by chemicals that activate the aryl hydrocarbon receptor (AhR). The strength of association between the chemical exposure and cancer has in this case been strengthened by studies that have shown that experimental exposure of animals to, for example, benzo(a)pyrene, a model activator of the AhR pathway, is associated with increased occurrence of tumors. The association between pollutant exposure and tumor incidence has been shown in such phylogenetically widely diverging organisms as molluscs and fish, indicating that xenobiotically induced carcinogenesis is a very basic response to aquatic contamination. The contaminants that activate the AhR pathway may cause carcinogenesis either directly or after transformation to more toxic intermediates in phase 1 of biotransformation (see section 9.1.1).

11.12 BEHAVIORAL EFFECTS

Behavior is the integrated output of the nervous system. Thus, there is a close connection between neurotoxicity and behavioral toxicity. However, neural and behavioral effects are often kept separate, because the workers in these fields and the methods commonly utilized are different. The most commonly used behavioral end points pertain to locomotion or feeding, although other, more specific, behaviors are occasionally also targeted. Commonly measured end points related to locomotion include the distance swum per unit time, the number of turns per unit time, the speed of burying, and the amount of time spent hidden/exposed. For burying animals, the location of the contaminant affects behavioral toxicity results markedly: if the sediment is the source of toxicity, the burying speed and time spent exposed are different from those observed if the source of toxicity is bulk water. Development of video and computer technology has markedly helped in defining quantitative

FIGURE 11.19 **Feeding behavior as an example of measurable behavioral responses to toxicants.**

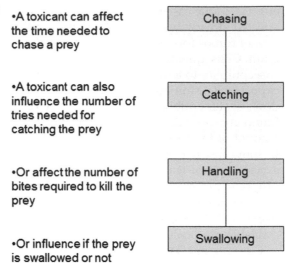

•A toxicant can affect the time needed to chase a prey

•A toxicant can also influence the number of tries needed for catching the prey

•Or affect the number of bites required to kill the prey

•Or influence if the prey is swallowed or not

locomotory end points. Figure 11.19 gives a schematic representation of behaviors associated with feeding, as an example of behaviors and end points associated with them that can be measured.

Relevant Literature and Cited References

Abele, D., Vazquez-Medina, P., Zenteno-Savin, T., 2012. Oxidative Stress in Aquatic Ecosystems. Wiley-Blackwell, Chichester, UK. 524 pp.

Alavi, S.M.H., Cosson, J., 2005. Sperm motility in fishes, (I) Effects of temperature and pH: A review. Cell. Biol. Int. 29, 101–110.

Alavi, S.M.H., Cosson, J., 2006. Sperm motility in fishes, (II) Effects of ions and osmolality: A review. Cell. Biol. Int. 30, 1–14.

Bantle, J.A., Fort, D.J., Dawson, D.A., 1988. Bridging the gap from short-term teratogenesis assays to human health hazard assessment by understanding common modes of teratogenic action (ASTM STP 1027). In: Cowgill, U.M., Williams, L.R. (Eds.). Aquatic Toxicology and Hazard Assessment, vol. 12. American Society for Testing and Materials, Philadelphia, PA, pp. 46–58.

Barata, C., Porte, C., Baird, D.J., 2004. Experimental designs to assess endocrine-disrupting effects in invertebrates: A review. Ecotoxicology 13, 511–517.

Bolognesi, C., Hayashi, M., 2011. Micronucleus assay in aquatic animals. Mutagenesis 26, 205–213.

Carvan, M.J., Goksøyr, A., Gallagher, E.P., Hahn, M.E., Larsson, D.G.J., 2007. Fish models in toxicology. Zebrafish 4, 9–20.

Chipman, J.K., Marsh, J.W., 1991. Bio-techniques for the detection of genetic toxicity in the aquatic environment. J. Biotechnol. 17, 199–208.

Debenest, T., Silvestre, J., Coste, M., Pinelli, E., 2010. Effects of pesticides on freshwater diatoms. Rev. Environ. Contam. Toxicol. 203, 87–103.

Denslow, N.D., Garcia-Reyero, N., Barber, D.S., 2007. Fish 'n' chips: The use of microarrays for aquatic toxicology. Mol. Biosyst. 3, 172–177.

Fent, K., Sumpter, J.P., 2011. Progress and promises in toxicogenomics in aquatic toxicology: Is technical innovation driving scientific innovation? Aquat. Toxicol. 105. Suppl, 25–39.

Froehlicher, M., Liedtke, A., Groh, K.J., Neuhauss, S.C., Segner, H., Eggen, R.I., 2009. Zebrafish (Danio rerio) neuromast: Promising biological endpoint linking developmental and toxicological studies. Aquat. Toxicol. 95, 307–319.

Fulton, M.H., Key, P.B., 2001. Acetylcholinesterase inhibition in estuarine fish and invertebrates as an indicator of organophosphorus insecticide exposure and effects. Environ. Toxicol. Chem. 20, 37–45.

Galloway, T.S., Depledge, M.H., 2001. Immunotoxicity in invertebrates: Measurement and ecological relevance. Ecotoxicology 10, 5–23.

Giattina, J.D., Garton, R.R., 1983. A review of the preference–avoidance response of fishes to aquatic contaminants. Residue Rev. 87, 43–90.

Gilbert, S.F., Epel, D., 2009. Ecological Developmental Biology: Integrating Epigenetics, Medicine, and Evolution. Sinauer Associates, Sunderland, MA.

Gracey, A.Y., Cossins, A.R., 2003. Application of microarray technology in environmental and comparative physiology. Annu. Rev. Physiol. 65, 231–259.

Guler, Y., Ford, A.T., 2010. Anti-depressants make amphipods see the light. Aquat. Toxicol. 99, 397–404.

Halliwell, B., Gutteridge, J.M.C., 2007. Free Radicals in Biology and Medicine, fourth ed. Oxford University Press, Oxford, UK.

Hellou, J., Ross, N.W., Moon, T.W., 2012. Glutathione, glutathione-S-transferase, and glutathione conjugates, complementary markers of oxidative stress in aquatic biota. Environ. Sci. Pollut. Res. Int. 19, 2007–2023.

Hines, A., Staff, F.J., Widdows, J., Compton, R.M., Falciani, F., Viant, M.R., 2010. Discovery of metabolic signatures for predicting whole organism toxicology. Toxicol. Sci. 115, 369–378.

Hinton, D.E., Kullman, S.W., Hardman, R.C., Volz, D.C., Chen, P.J., Carney, M., Bencic, D.C., 2005. Resolving mechanisms of toxicity while pursuing ecotoxicological relevance? Mar. Pollut. Bull. 51, 635–648.

Högstrand, C., Kille, P. (Eds.), 2008. Comparative toxicogenomics. Adv. Exp. Biol. 2, 1–329.

Ju, Z., Wells, M.C., Walter, R.B., 2007. DNA microarray technology in toxicogenomics of aquatic models: Methods and applications. Comp. Biochem. Physiol. C. Toxicol. Pharmacol. 145, 5–14.

Kassahn, K.S., Caley, M.J., Ward, A.C., Connolly, A.R., Stone, G., Crozier, R.H., 2007. Heterologous microarray experiments used to identify the early gene response to heat stress in a coral reef fish. Mol. Ecol. 16, 1749–1763.

Kloas, W., Urbatzka, R., Opitz, R., Wurtz, S., Behrends, T., Hermelink, B., Hofmann, F., Jagnytsch, O., Kroupova, H., Lorenz, C., Neumann, N., Pietsch, C., Trubiroha, A., Van, B.C., Wiedemann, C., Lutz, I., 2009. Endocrine disruption in aquatic vertebrates. Ann. N. Y. Acad. Sci. 1163, 187–200.

Krumschnabel, G., Podrabsky, J.E., 2009. Fish as model systems for the study of vertebrate apoptosis. Apoptosis 14, 1–21.

La, D.K., Swenberg, J.A., 1996. DNA adducts: Biological markers of exposure and potential applications to risk assessment. Mutat. Res. Rev. Gen. Toxicol. 365, 129–146.

Lafont, R., Mathieu, M., 2007. Steroids in aquatic invertebrates. Ecotoxicology 16, 109–130.

Lannoo, M., 2008. Malformed Frogs: The Collapse of Aquatic Ecosystems. University of California Press, Berkeley, CA.

LeBlanc, G.A., 2007. Crustacean endocrine toxicology: A review. Ecotoxicology 16, 61–81.

Lushchak, V.I., 2011. Environmentally induced oxidative stress in aquatic animals. Aquat. Toxicol. 101, 13–30.

Martyniuk, C.J., Denslow, N.D., 2012. Exploring androgen-regulated pathways in teleost fish using transcriptomics and proteomics. Integr. Comp. Biol. 52, 695–704.

Martyniuk, C.J., Popesku, J.T., Chown, B., Denslow, N.D., Trudeau, V.L., 2012. Quantitative proteomics in teleost fish: Insights and challenges for neuroendocrine and neurotoxicology research. Gen. Comp. Endocrinol. 176, 314–320.

Meyer, J.N., Leung, M.C., Rooney, J.P., Sendoel, A., Hengartner, M.O., Kisby, G.E., Bess, A.S., 2013. Mitochondria as a target of environmental toxicants. Toxicol. Sci. 134, 1–17.

Miracle, A.L., Ankley, G.T., 2005. Ecotoxicogenomics: Linkages between exposure and effects in assessing risks of aquatic contaminants to fish. Reprod. Toxicol. 19, 321–326.

Narahashi, T., 1987. Nerve membrane ion channels as the target site of environmental toxicants. Environ. Health Perspect. 71, 25–29.

Nikinmaa, M., Rytkonen, K.T., 2011. Functional genomics in aquatic toxicology: Do not forget the function. Aquat. Toxicol. 105. Suppl, 16–24.

Ohe, T., Watanabe, T., Wakabayashi, K., 2004. Mutagens in surface waters: A review. Mutat. Res. 567, 109–149.

Paskova, V., Hilscherova, K., Blaha, L., 2011. Teratogenicity and embryotoxicity in aquatic organisms after pesticide exposure and the role of oxidative stress. Rev. Environ. Contam. Toxicol. 211, 25–61.

Pina, B., Barata, C., 2011. A genomic and ecotoxicological perspective of DNA array studies in aquatic environmental risk assessment. Aquat. Toxicol. 105. Suppl, 40–49.

Porte, C., Janer, G., Lorusso, L.C., Ortiz-Zarragoitia, M., Cajaraville, M.P., Fossi, M.C., Canesi, L., 2006. Endocrine disruptors in marine organisms: Approaches and perspectives. Comp. Biochem. Physiol. C. Toxicol. Pharmacol. 143, 303–315.

Pounds, J.G., 1990. The role of cell calcium in current approaches to toxicology. Environ. Health Perspect. 84, 7–15.

Robertson, D.G., 2005. Metabonomics in toxicology: A review. Toxicol. Sci. 85, 809–822.

Rotchell, J.M., Ostrander, G.K., 2003. Molecular markers of endocrine disruption in aquatic organisms. J. Toxicol. Environ. Health B. Crit. Rev. 6, 453–496.

Sanchez, B.C., Ralston-Hooper, K., Sepulveda, M.S., 2011. Review of recent proteomic applications in aquatic toxicology. Environ. Toxicol. Chem. 30, 274–282.

Scott, G.R., Sloman, K.A., 2004. The effects of environmental pollutants on complex fish behaviour: Integrating behavioural and physiological indicators of toxicity. Aquat. Toxicol. 68, 369–392.

Segner, H., 2009. Zebrafish (*Danio rerio*) as a model organism for investigating endocrine disruption. Comp. Biochem. Physiol. C. 149, 187–195.

Sikkema, J., De Bont, J.A.M., Poolman, B., 1995. Mechanisms of membrane toxicity of hydrocarbons. Microbiol. Rev. 59, 201–222.

Silva, L.J., Lino, C.M., Meisel, L.M., Pena, A., 2012. Selective serotonin re-uptake inhibitors (SSRIs) in the aquatic environment: An ecopharmacovigilance approach. Sci. Total. Environ. 437, 185–195.

Snape, J.R., Maund, S.J., Pickford, D.B., Hutchinson, T.H., 2004. Ecotoxicogenomics: The challenge of integrating genomics into aquatic and terrestrial ecotoxicology. Aquat. Toxicol. 67, 143–154.

Soffker, M., Tyler, C.R., 2012. Endocrine-disrupting chemicals and sexual behaviors in fish: A critical review on effects and possible consequences. Crit. Rev. Toxicol. 42, 653–668.

Sumpter, J.P., 1998. Xenoendorine disrupters: Environmental impacts. Toxicol. Lett. 102-103, 337–342.

Taylor, N.S., Weber, R.J., White, T.A., Viant, M.R., 2010. Discriminating between different acute chemical toxicities via changes in the daphnid metabolome. Toxicol. Sci. 118, 307–317.

Taylor, R.C., Cullen, S.P., Martin, S.J., 2008. Apoptosis: Controlled demolition at the cellular level. Nat. Rev. Mol. Cell. Biol. 9, 231–241.

Valavanidis, A., Vlahogianni, T., Dassenakis, M., Scoullos, M., 2006. Molecular biomarkers of oxidative stress in aquatic organisms in relation to toxic environmental pollutants. Ecotoxicol. Environ. Saf. 64, 178–189.

Van Beneden, R.J., 1997. Environmental effects and aquatic organisms: Investigations of molecular mechanisms of carcinogenesis. Environ. Health Perspect. 105 (Suppl. 3), 669–674.

Vasconcelos, V., Azevedo, J., Silva, M., Ramos, V., 2010. Effects of marine toxins on the reproduction and early stages of development of aquatic organisms. Mar. Drugs 8, 59–79.

Viant, M.R., 2008. Environmental metabolomics using 1H-NMR spectroscopy. Methods Mol. Biol. 410, 137–150.

Weis, J.S., Weis, P., 1987. Pollutants as developmental toxicants in aquatic organisms. Environ. Health Perspect. 71, 77–85.

Wester, P.W., Vethaak, A.D., van Muiswinkel, W.B., 1994. Fish as biomarkers in immunotoxicology. Toxicology 86, 213–223.

Williams, T.D., Turan, N., Diab, A.M., Wu, H., Mackenzie, C., Bartie, K.L., Hrydziuszko, O., Lyons, B.P., Stentiford, G.D., Herbert, J.M., Abraham, J.K., Katsiadaki, I., Leaver, M.J., Taggart, J.B., George, S.G., Viant, M.R., Chipman, K.J., Falciani, F., 2011. Towards a system-level understanding of non-model organisms sampled from the environment: A network biology approach. PLoS Comput. Biol. 7, e1002126.

Yang, G., Kille, P., Ford, A.T., 2008. Infertility in a marine crustacean: Have we been ignoring pollution impacts on male invertebrates? Aquat. Toxicol. 88, 81–87.

Bioindicators and Biomarkers

Abstract

Bioindicator species are used in biomonitoring contaminant exposure. In order to evaluate how contamination exists spatially, sessile species are the best. In addition, a good bioindicator species is common and easily sampled, presenting some responses to toxicants (biomarker responses) that can be reliably measured and show concentration dependence. The suitability of bioindicator species varies markedly between aquatic bodies, as the sensitivities of organisms to different contaminants vary markedly. The commonly used model organisms are good bioindicator species only when they are common in the natural environment studied. Biomarkers of exposure indicate that the species has been exposed to a toxicant. Good examples are mRNAs, which show that exposure affects transcription, but do not give information on the effects of toxicants on the function of organisms. This is done by biomarkers of effect, which usually measure protein activities. The biomarker responses can be markedly affected by the presence of other compounds in the environment, by natural environmental stresses, and by the length of exposure. Effect–biomarker responses translate to ecosystem effects if the measured parameter affects the fitness of the organism studied.

Keywords: Mussel Watch; biomonitoring species; biological indices; hydra; algae; contaminant trends; biomarker of exposure; biomarker of effect; biomarker of susceptibility; early warning signal; sublethal effect; exposure time; integrated biomarker suite.

12.1 BIOINDICATORS

Bioindicators (biomonitoring species) can be defined as species or groups of species that are used to indicate adverse effects of contamination. Bioindicator species are usually different from the model species used for toxicological research (Chapter 5), since the model species are often not species found in the natural environments monitored. The adverse effects

FIGURE 12.1 **Bioindicator species are species found in the natural environment that are used in biomonitoring.** (1) Pelagic species such as the shark monitor mainly the aquatic environment. Green plants with roots and shellfish get toxicants both from the aquatic phase and sediment. (2) The usefulness of pelagic organisms as bioindicators depends on their immigration to the area and emigration from the area. In addition, selective mortality because of toxicants affects the measured values from bioindicators. (3) Sessile organisms such as plants, molluscs, and sea anemones are highly suitable for evaluating local contamination.

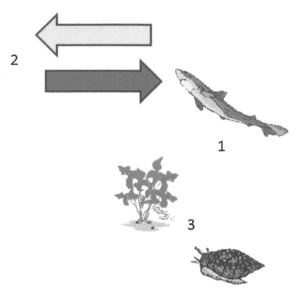

can be measurable responses in the organisms, or changes in the number of specimens or proportional abundance of the species in communities. The proportional abundances of different indicator species (biodiversity, species richness) are used to calculate different biological indices, as discussed in Chapter 18. A good bioindicator species is common in a variety of environments. This enables it to be used in estimation of ecosystem health in various instances. It is also tolerant to toxicants, but has an easily measured property that is sensitive to a variety of toxicants. Further, the immigration to or emigration from the studied environment does not occur. The aspects of a good bioindicator species are given in Figure 12.1. The different properties of a good bioindicator species cannot easily be found in a single organism. Rather, quite often, the population density of a species is used as an indicator, or, instead of evaluating the density of a single species, different biological indices, variations in the abundances of species groups, are used to indicate the contamination of the environment.

One specific group of organisms that has been advocated to be used as bioindicators of the health of aquatic ecosystems is mussels. The so-called Mussel Watch was started in the USA in 1986, and has expanded so that, in addition to US coastal and Great Lakes waters, mussels are used for monitoring aquatic contamination throughout the world. The Mussel Watch project was developed to analyze contaminant trends in sediments and bivalve tissues (over 100 organic and inorganic contaminants). The contaminants quantified include polyaromatic hydrocarbons (PAHs), polychlorinated biphenyls (PCBs), the insecticide DDT and its metabolites, tributyl tin (TBT) and its metabolites, chlorinated pesticides, and toxic trace elements. While mussels are in some respects good bioindicator organisms as they contain species with wide distribution and as they are quite sessile whereby the accumulation of compounds in them represents local contamination, the fact that they are shelled can influence chemical

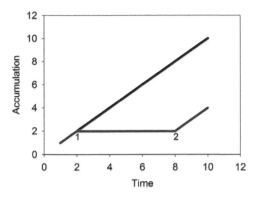

FIGURE 12.2 **The effects of shell closure on toxicant accumulation in bivalves.** (A) The shell is open throughout exposure (black line). (B) The bivalve senses the exposure to a harmful toxicant and closes the shell (red line). From the point of shell closure (1), no accumulation occurs until the animal needs to open the shell (2) to obtain oxygen. The amount of toxicant taken up in the latter case is only a fraction (in this case 40%) of that in the former case.

uptake. Mussels are capable of completely closing their shell, and they can survive up to several months without opening the shell at all to obtain oxygen. Different chemicals affect shell closure in different ways: the animals do not react to the presence of some chemicals at all, while others cause shell closure (Figure 12.2). The uptake of the first set of chemicals can reliably be followed using mussels, while contamination by the latter may remain unnoticed (if the toxicant is effectively diluted before the shell must be opened to obtain oxygen). Thus, sessile animals without a shell, e.g. sea anemones and hydras, macroscopic algae, and aquatic green plants, would actually be preferable for studying the bioaccumulation of toxicants. Regardless of the organism used for bioaccumulation studies, one must bear in mind that studying the effects of toxicants on organisms needs to be done with the organism type(s) that are most likely to be affected by contamination, in order to assess the possible ecosystem effects. Also, the type of organism influences the contamination type that is evaluated. In the case of pelagic fish, invertebrates, and algae, only aquatic contamination is targeted (+ contamination from food). In the case of benthic organisms, mainly sediment toxicity is evaluated. In the case of rooted aquatic plants and algae, both sediments and the aquatic phase are the source of toxic chemicals.

12.2 BIOMARKERS

A biomarker is a measurable trait in an organism that responds to a toxicant. Thus, a biomarker is actually a disturbance of a normal of function of an organism. Because of this, the biomarker field needs an understanding of the normal physiology of organisms, how this is disturbed by contaminants, and what consequences the disturbances of function may have in terms of the fitness of an organism. Examples of biomarkers are given in Table 12.1. Ideally, biomarker results should enable one to evaluate a potential risk to populations, communities, or ecosystems. They are used in monitoring possible effects of environmental contamination, and give "early warning signals" before irreversible damage occurs (sublethal effects). The properties of a good biomarker are:

TABLE 12.1 Biomarker Responses of Animals Commonly Used in Aquatic Toxicology

Biomarker	Biomarker Type	Remarks
ALA-D activity	Exposure and effect	Δ-aminolevulinic acid dehydratase catalyzes one enzymatic reaction in heme production. Only lead exposure is known to inhibit the reaction. Thus, observed inhibition indicates that the organism has been exposed to lead and that the exposure causes problems in oxygen transport and in all the reactions where heme groups are involved (i.e. all cytochrome enzymes)
cDNA (complementary DNA) microarrays	Exposure	As a large number of transcripts are targeted, virtually all contaminants give different signatures. However, it must be borne in mind that the transcript information does not give the mode of action, since one does not know if any toxicant-responsive gene products are formed and their activities are changed as a response to the environmental change
Bile composition	Exposure	The exposure to specific organic toxicants can be evaluated from their appearance, their breakdown, or conjugation products in the bile of exposed animals
Spiggin	Effect	One of the few biomarkers for the detection of androgenic effects in fish. The protein is the glue protein that is used by stickleback males to build nests where the offspring are reared. Androgenic contamination affects spiggin levels in males. In heavily contaminated sites, can even be secreted by females
Vitellogenin	Effect	The protein is produced in the livers of fish as a precursor of vitellins that form the yolk proteins of eggs. Specific antibodies for the detection of the protein have been developed. In an environment contaminated by estrogen agonists, males also often produce the compound. The production is species-specific, and although vitellogenin production in invertebrates has been used as biomarker of estrogenic effects, the actual regulation of vitellogenin production in different invertebrates is poorly known
Acetylcholinesterase activity	Effect	The activity of the enzyme indicates the presence of contaminants affecting synaptic function
Retinol profile	Exposure	Retinol is vitamin A. Organochlorine compounds affect the proportions of different retinols. Since the exact functions of vitamin A in animals are poorly known, the effects of an altered profile are likewise not clear
Porphyrin profiles	Exposure	Porphyrins are integral parts of heme (of globins and cytochromes). Different porphyrins can be characterized by liquid chromatography. The profiles change as a response to organochlorine contamination

Biomarker	Type	Description
Mixed-function oxidases	Exposure	Mixed-function oxidases are the major enzymes of phase 1 in biotransformation. They usually contain cytochrome P450. In the case that only mRNA of cyp enzymes is determined, the increase mostly indicates contamination by PAHs. Antibodies against some major cytochrome P450s in fish have also been generated.
EROD (ethoxyresorufin-O-deethylase) activity	Effect	The enzyme activity is used to show changes in the activity of phase 1 of biotransformation, mainly by PAHs and other compounds that activate the aryl hydrocarbon receptor pathway
Metallothionein levels	Exposure	Levels are measured especially to indicate exposure to cadmium, copper, zinc, and mercury (and other metals). Problems in interpretation are caused by the facts that metallothionein induction is not always observed; that there are marked differences in induction between different species; that in addition to being induced by metals, metallothioneins are involved in redox regulation; and that their levels and induction are cell-cycle-stage specific.
DNA adducts	Exposure	DNA repair makes interpreting the results difficult. DNA adducts are formed especially by bulky aromatic compounds
Lipid peroxidation	Effect	Measured as MDA (malonyl dialdehyde), one of the final products of lipid peroxidation. The most common method used is evaluating TBARS (thiobarbituric acid reactive substances) levels, the most important of which is MDA. All conditions and chemicals that cause oxidative stress cause an increase in TBARS
Comet assay	Effect	Indicates the formation of DNA fragments. Caused by various genotoxicants
Micronucleus test	Effect	Caused by various genotoxicants. Results either from disturbances in the function of the mitotic spindle during cell division or fragmentation of DNA so that daughter cells have a main nucleus and a smaller micronucleus
TUNEL assay	Effect	The TUNEL assay is a common method for determining DNA fragmentation characteristic of apoptotic cell death. Many contaminants cause apoptosis, and so the specificity of the response is low
Stress protein (heat shock protein, HSP) levels	Effect	Stress protein levels increase after all treatments that affect the normal three-dimensional structure of proteins. Thus, virtually all toxicants can affect HSP levels

The biomarkers are arranged according to their specificity. In addition to the presented biomarkers, many other are used, but typically either their specificity is low (e.g. if redox parameters are used as biomarkers, many compounds, from metals to PAHs, can cause changes in them) or what the response actually means for the function of organisms is not known.

1. It is easily measured.
2. The measurements are fast and cheap.
3. The measurement is specific to a toxicant type.
4. The response shows a concentration–response (dose–response) relationship.

The fact that very often a concentration–response relationship is not observed is schematically illustrated in Figure 12.3. It must always be remembered that different organisms vary markedly in their sensitivity to different contaminants. Thus, biomarker effects have the best ecological relevance when they can be combined with knowledge of the sensitivity of the species to contamination. As is clear from the above consideration of the properties of good biomarkers, the situation is the same as for bioindicator species: they can seldom be found in a single response. The properties of biomarkers can further be influenced not only by the chemical(s) that they are intended to monitor, but by interactions with other chemicals present in the environment (Chapter 13), and the prevailing environmental conditions (Chapter 16) can markedly influence biomarker responses, as schematically illustrated in Figure 12.4. Biomarkers can be divided into biomarkers of exposure and biomarkers of effect. In addition, one can define a biomarker of susceptibility: this is an inherent or acquired ability of an organism to respond to exposure to a specific substance. In a way it comes before the effect biomarker, as it indicates the possibility that chemical exposure has an effect. A biomarker response need not be associated with the toxic effects, the only requirement is that a change is observed (in which case the change is a biomarker of exposure). Consequently, any transcriptional effect can be such a biomarker even if a change in transcription is not later reflected in protein activity. Furthermore, since a multitude of genes can be evaluated at the same time (see section 11.1), the measurements are fast. Preparing samples for measurements is also simple and samples can be saved at room temperature in the RNA-stabilization agent RNAlater®. So, in many ways, microarray chips are a good basis for biomarker development. Currently, the major drawback of all the 'omics work in the biomarker field is that it requires expensive equipment or sample processing, which are not universally available.

An important consideration in the biomarker field is the exposure time, and the time it takes for an organism to respond. This brings together, as discussed above, the suitability of bioindicators and biomarkers. Quite often, biomarkers are used for biomonitoring

FIGURE 12.3 **The reason why biomarker responses do not follow a concentration–response relationship.** (1) Initially, an increasing toxicant concentration causes a linear increase in the biomarker response. The increase may be due both to an increase in protein production and to an increase in the maximal activity of the protein(s) involved in the response. (2) The increase slows down and a maximal response is reached, whereafter (3) the chemical becomes acutely toxic to the response and may decrease the maximal activity of the protein molecule and/or decrease the production of the protein (possibly both transcription and translation are inhibited).

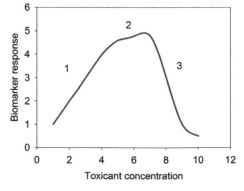

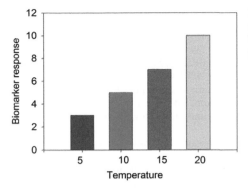

FIGURE 12.4 **A hypothetical example of how a natural environmental variable (temperature) affects a biomarker response.** With an increase in temperature, the value of the response increases markedly.

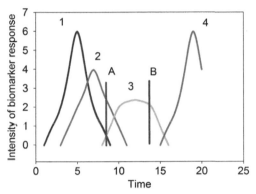

FIGURE 12.5 **A schematic example of temporal differences in biomarker responses to a toxicant.** The y axis gives the relative change in the biomarker value in comparison to the control. At 0 the control values and those measured from exposed organisms are the same. The different biomarkers (1–4) each have a unique time profile of responses. If sampling can be done only at one point, biomarkers 1–3 are all altered in comparison to the control at time point A, but only biomarker 3 at time point B. All the different biomarker responses cannot be measured at a single time point. Even at time point A, where three biomarkers show alterations, no values are close to the maximal response of any of the biomarkers.

potential effects of environmental pollution in natural environments. To enable this to be done, the responses to contamination must be apparent in a time frame in which contamination is expected to have occurred. Alternatively, there must be a difference in biomarker response between putatively affected and unaffected areas. As discussed in Chapter 14, an important component of toxic responses is that they differ temporally. In view of this, if a biomarker sample is taken at the wrong time, no response may be obtained even if the toxicant would affect the organism and possibly the ecosystem. For this reason, biomarkers should be determined at more than one or two time points. In addition, when biomonitoring natural environments the contamination history is usually unclear. Consequently, it is possible that some biomarker responses usually associated with toxicants are no longer seen (see Figure 12.5).

A single biomarker response is usually not adequate, and to improve the possibilities for predicting the contamination of the environment and its consequences on organisms, a suite of biomarkers should be used. To increase the possibility of choosing the right biomarkers, likely contaminants in the environment should be evaluated, and possibly measured, and the integrated biomarker suite chosen based on this information. The use of effect biomarkers integrates the bioavailability and excretion of a compounds, to gain knowledge about the effective dose in an organism.

Relevant Literature and Cited References

Augusto, S., Maguas, C., Branquinho, C., 2013. Guidelines for biomonitoring persistent organic pollutants (POPs), using lichens and aquatic mosses: A review. Environ. Pollut. 180, 330–338.

Bebianno, M.J., Geret, F., Hoarau, P., Serafim, M.A., Coelho, M.R., Gnassia-Barelli, M., Romeo, M., 2004. Biomarkers in *Ruditapes decussatus*: A potential bioindicator species. Biomarkers 9, 305–330.

Bergek, S., Ma, Q., Vetemaa, M., Franzen, F., Appelberg, M., 2012. From individuals to populations: Impacts of environmental pollution on natural eelpout populations. Ecotoxicol. Environ. Saf. 79, 1–12.

Brain, R.A., Cedergreen, N., 2009. Biomarkers in aquatic plants: Selection and utility. Rev. Environ. Contam. Toxicol. 198, 49–109.

Campos, A., Tedesco, S., Vasconcelos, V., Cristobal, S., 2012. Proteomic research in bivalves: Towards the identification of molecular markers of aquatic pollution. J. Proteomics 75, 4346–4359.

Conti, M.E. (Ed.), 2008. Biological Monitoring: Theory and Applications. WIT Press, Southampton, UK.

Domingues, I., Agra, A.R., Monaghan, K., Soares, A.M., Nogueira, A.J., 2010. Cholinesterase and glutathione-S-transferase activities in freshwater invertebrates as biomarkers to assess pesticide contamination. Environ. Toxicol. Chem. 29, 5–18.

Ferrat, L., Pergent-Martini, C., Romeo, M., 2003. Assessment of the use of biomarkers in aquatic plants for the evaluation of environmental quality: Application to seagrasses. Aquat. Toxicol. 65, 187–204.

Forbes, V.E. (Ed.), 1998. Genetics and Ecotoxicology. CRC Press, Boca Raton, FL.

Ph, Garrigues, Barth, H., Walker, C.H., Narbonne, J.-F. (Eds.), 2001. Biomarkers in Marine Organisms: A Practical Approach. Elsevier, Amsterdam, The Netherlands.

Graczyk, T.K., Conn, D.B., 2008. Molecular markers and sentinel organisms for environmental monitoring. Parasite 15, 458–462.

Hyne, R.V., Maher, W.A., 2003. Invertebrate biomarkers: Links to toxicosis that predict population decline. Ecotoxicol. Environ. Saf. 54, 366–374.

Lewis, C., Ford, A.T., 2012. Infertility in male aquatic invertebrates: A review. Aquat. Toxicol. 120-121, 79–89.

Mao, H., Wang, D.H., Yang, W.X., 2012. The involvement of metallothionein in the development of aquatic invertebrates. Aquat. Toxicol. 110-111, 208–213.

Marin, M.G., Matozzo, V., 2004. Vitellogenin induction as a biomarker of exposure to estrogenic compounds in aquatic environments. Mar. Pollut. Bull. 48, 835–839.

Markert, B.A., Breure, A.M., Zechmeister, H.G., 2003. Bioindicators and Biomonitors. Gulf Professional, Burlington, MA.

Matozzo, V., Gagne, F., Marin, M.G., Ricciardi, F., Blaise, C., 2008. Vitellogenin as a biomarker of exposure to estrogenic compounds in aquatic invertebrates: A review. Environ. Int. 34, 531–545.

Miracle, A.L., Toth, G.P., Lattier, D.L., 2003. The path from molecular indicators of exposure to describing dynamic biological systems in an aquatic organism: Microarrays and the fathead minnow. Ecotoxicology 12, 457–462.

Nesatyy, V.J., Suter, M.J., 2007. Proteomics for the analysis of environmental stress responses in organisms. Environ. Sci. Technol. 41, 6891–6900.

Ostrander, G.K. (Ed.), 1996. Techniques in Aquatic Toxicology. CRC Press, Boca Raton, FL.

Ostrander, G.K. (Ed.), 2005. Techniques in Aquatic Toxicology, vol. 2. CRC Press, Boca Raton, FL.

Palos, L.M., Bigot, A., Aubert, D., Hohweyer, J., Favennec, L., Villena, I., Geffard, A., 2013. Protozoa interaction with aquatic invertebrate: Interest for watercourses biomonitoring. Environ. Sci. Pollut. Res. Int. 20, 778–789.

Parrott, J.L., McMaster, M.E., Hewitt, L.M., 2006. A decade of research on the environmental impacts of pulp and paper mill effluents in Canada: Development and application of fish bioassays. J. Toxicol. Environ. Health B. Crit. Rev. 9, 297–317.

Peakall, D.B., 1992. Animal Biomarkers as Pollution Indicators. Chapman and Hall, London.

Quinn, B., Gagne, F., Blaise, C., 2012. Hydra, a model system for environmental studies. Int. J. Dev. Biol. 56, 613–625.

Stegeman, J.J., Lech, J.J., 1991. Cytochrome P-450 monooxygenase systems in aquatic species: Carcinogen metabolism and biomarkers for carcinogen and pollutant exposure. Environ. Health Perspect. 90, 101–109.

Sumpter, J.P., Jobling, S., 1995. Vitellogenesis as a biomarker for estrogenic contamination of the aquatic environment. Environ. Health Perspect. 103 (Suppl. 7), 173–178.

Torres, M.A., Barros, M.P., Campos, S.C., Pinto, E., Rajamani, S., Sayre, R.T., Colepicolo, P., 2008. Biochemical biomarkers in algae and marine pollution: A review. Ecotoxicol. Environ. Saf. 71, 1–15.

Valavanidis, A., Vlahogianni, T., Dassenakis, M., Scoullos, M., 2006. Molecular biomarkers of oxidative stress in aquatic organisms in relation to toxic environmental pollutants. Ecotoxicol. Environ. Saf. 64, 178–189.

Zelikoff, J.T., Raymond, A., Carlson, E., Li, Y., Beaman, J.R., Anderson, M., 2000. Biomarkers of immunotoxicity in fish: From the lab to the ocean. Toxicol. Lett. 112-113, 325–331.

Zhou, Q., Zhang, J., Fu, J., Shi, J., Jiang, G., 2008. Biomonitoring: An appealing tool for assessment of metal pollution in the aquatic ecosystem. Anal. Chim. Acta. 606, 135–150.

Interactions between Chemicals

Abstract

In natural environments, contaminants occur in cocktails that may have no agonistic or antagonistic interactions. If the compounds have no interactions, the different toxicants do not influence the properties of each other's sites of action or any of the pathways involved in generating the response. In the case of agonism, the affinity of a chemical for its site of action is increased by another chemical, or the detoxification reactions are slowed down or the level of more toxic intermediates is increased by the second chemical. Antagonism is opposite to agonism, whereby either the affinity of a toxicant for its site of action is reduced, the detoxification of contamination speeded up, or levels of more toxic metabolites reduced by the additional chemical. Although complex cocktails are characteristic of natural environments, virtually no studies are available about agonism or antagonism in natural environments.

Keywords: cocktail effects; concentration addition; dioxin equivalent; toxic equivalence; independent action; agonism; potentiation; synergism; daughter compound toxicity; antagonism; inhibition; allosteric regulation; delayed toxicity; synergism ratio.

13.1 LACK OF INTERACTING EFFECTS (ADDITIVE TOXICITY)

In natural environments, chemicals occur in mixtures (cocktails). Because of this, interactive effects between chemicals always need to be considered. In the simplest case, chemicals do not affect each other's toxicity. In this case, toxicity follows a straight line (Figures 13.1 and 13.2) that is the sum of concentration of all chemicals in the mixture. This type of behavior

FIGURE 13.1   **A scheme for toxic interactions of a binary mixture of toxicants.** A value of 0 on the y axis indicates that the toxicity is the same as the pure toxicant A or B. Positive values indicate agonism (synergism, potentiation) and negative values antagonism (inhibition). (1) The two chemicals exhibit potentiation, increasing each other's toxicity. (2) The chemicals do not affect each other's toxicity. (3) The chemicals decrease each other's toxicity, thus showing antagonism (inhibition). In addition to toxicants showing reciprocal effects, it is possible that one of the toxicants affects the function of the other, but not vice versa.

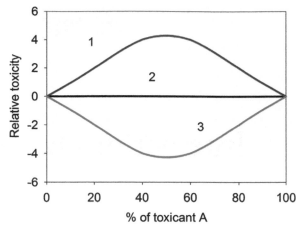

FIGURE 13.2   **The influence of potentiation and antagonism by another compound on the toxicity of a chemical.** The concentration of the chemical is increased while that of the interacting compound is constant. (A) The effect increases linearly with toxicant concentration. (B) The increase in effect is nonlinear with concentration. The black line indicates no interaction between the chemicals, the red line potentiation, and the green line inhibition.

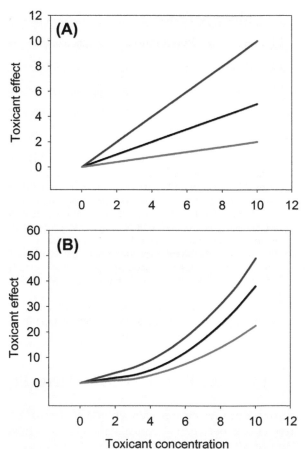

is typical when the compounds have the same site of action but do not affect the properties of the site. When the toxicity is calculated, one often uses the concentration addition (CA) model. Naturally, the toxicities of individual similarly acting compounds vary markedly. Perhaps the most common case of combining toxicities of individual similarly acting compounds concerns polychlorinated biphenyls (PCBs), furans, and dioxins. These compounds commonly act via the aryl hydrocarbon receptor (AhR) pathway, although other effect routes have also been described, especially for PCBs. Because toxicities of compounds vary markedly, toxicities can given as dioxin equivalents (toxicity of compound/toxicity of the dioxin TCDD) or, more generally, toxic equivalences (TEQs; or toxic equivalence factor, TEF). Often, only concentrations of toxicants that have a demonstrable effect alone are taken as components of a mixture. Very small concentrations are occasionally overlooked, but this should be avoided if the chemicals have the same site of action, as the combined concentrations may be adequate to cause toxicity. In addition to compounds acting via exactly the same site without affecting the properties of the site, compounds with independent actions (IA) may also exert toxicity that is directly additive, depending on the concentration. This requires that the toxic effects do not overlap in any way, e.g. the toxicant action does not affect the three-dimensional structure of the active site in the protein that is the target of the second toxicant.

13.2 AGONISM (POTENTIATION, SYNERGISM)

Agonism (potentiation, synergism) means that a compound increases the toxicity of a second compound (Figures 13.1 and 13.2). There are several ways that this can happen. The first and simplest one is that a chemical affects the three-dimensional structure of a protein in such a way that its affinity for the second chemical (and consequently its toxic actions) is increased. This is an example of allosteric regulation of protein function, which is a major way by which functions are regulated biochemically (Figure 13.3). A second possibility for agonism is that a chemical affects the biochemical pathway involved in the toxic action of a second chemical in such a way that the overall toxic effect is increased. An example of this can be given from endocrine disruption. A chemical increases aromatase activity, i.e. the interconversion between androgens and estrogens is facilitated. The toxic action of a second chemical is limited by such interconversion, i.e. by aromatase activity. As the first chemical increases aromatase activity, the limitation for the effect of the second chemical is relieved, whereby the overall toxicity of the second chemical is potentiated (for the general principle of such pathway effects, see Figure 13.4). The third possibility for potentiation of the toxic action of a compound is by inhibition of its detoxification (for the general principle, see Figure 13.5). If, for example, a chemical inhibits an enzyme of the AhR-dependent detoxification pathway, a toxicant that is normally metabolized via this pathway will have an increased effect. Fourth, occasionally the metabolites are more toxic than the parent compound. Examples of such daughter compounds have been found in phase 1 of detoxification. If the stability of this more toxic daughter compound is increased by the inhibition of an enzyme catalyzing its breakdown because of the second toxicant, the toxicity will increase agonistically. Similarly, if the second toxicant increases the activity of an enzyme catalyzing the formation of the more toxic daughter compound, its steady-state level will increase and be seen as potentiation of

FIGURE 13.3 **Allosteric potentiation of toxicant effect.** (1) The first toxicant (rectangle) induces the protein activity by binding to the protein (green). The second toxicant (red) binds to a different site of protein compared with the first. (2) This enables the first toxicant to bind to the affected protein with increased affinity, whereby the protein activity is increased compared to situation 1 (indicated with a change of protein color from green to blue).

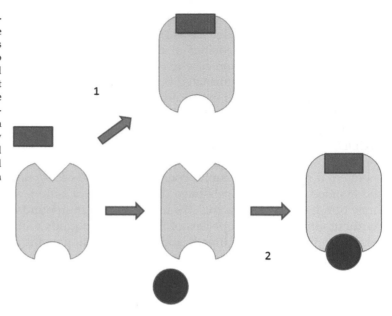

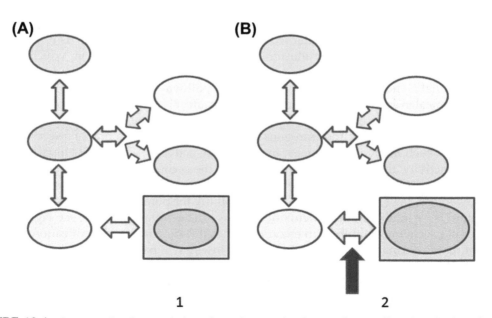

FIGURE 13.4 **An example of potentiation of a toxicant action by a pathway effect.** (A) The first chemical exerts its action mainly via an estrogenic effect (1). (B) The second chemical (red arrow; 2) increases the activity of aromatase, whereby the possibility for estrogen formation is increased and thus the toxicity of first chemical potentiated.

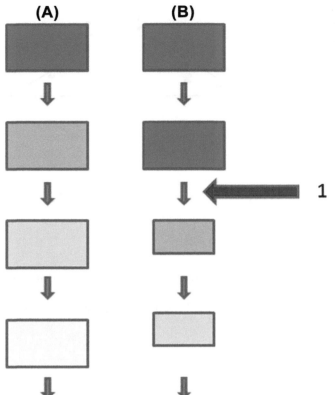

FIGURE 13.5 **Potentiation of toxicity by inhibition of detoxification.** (A) Normally, detoxification reactions decrease the toxicity of organic chemicals (shown in the figure as a progression from darker to lighter blue). (B) The second toxicant (red arrow; 1) inhibits detoxification, and as a result the toxicity of the first chemical increases.

toxicity. Fifth, in principle, a second toxicant can increase the production of harmful small molecules such as reactive oxygen species (ROS) during the detoxification of the first toxicant. Since ROS are toxic in their own right, an overall potentiation of toxicity follows from their increased production. From the above points, it is clear that understanding the possibilities of potentiation at the cellular level requires deep knowledge of cell physiology in order to take into account the different possible steps where potentiation may occur. This is all the more important as the different pathways leading to the potentiation of toxicant action may be different in different groups of organisms, or, even if the overall pathway of toxicant interactions is the same, the affinities of the different steps may be different.

Potentiation of toxic effects may also occur between seemingly unrelated phenomena. These are normally not considered under agonism, but the reader should be aware of the possibility of such interactions. For example, a toxicant increases the activity of an animal, increasing the use of lipid stores in energy production. This will increase the level of circulating lipid-soluble toxicants, whereby their toxic effects will also increase, although the total amount of toxicant in the body has not changed. Thus, one toxicant (affecting energy consumption) increases the toxic effects of another (which has earlier been to a great extent stored in lipid tissue and is, consequently, inert). However, this is normally not considered as potentiation of toxicity, but delayed toxicity.

FIGURE 13.6 **To be certain of potentiation or inhibition of toxicant action by a second toxicant, one needs to have enough measurements.** From the figure, it is clear that the number of determinations (red ellipses for supposed potentiation and green ellipses for inhibition) is not adequate to conclude that potentiation or inhibition has taken place, although the means would suggest this to be the case.

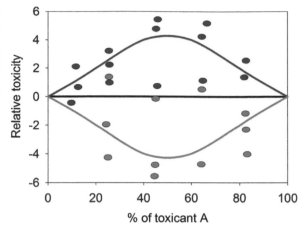

To be certain that the observed results truly show a potentiation of toxicity, and are not only an effect of errors accumulated in the measurements, an adequate number of samples must be analyzed (Figure 13.6). A rough "rule of the thumb" is that when the synergism ratio (SR) is more than 2, the phenomenon is potentiation:

$$SR = \frac{MEC_{chemical1}}{MEC_{chemical1 + chemical2}},$$

(13.1)

where MEC is median effective concentration (could also be MLC (median lethal concentration) or MLD (median lethal dose)).

13.3 ANTAGONISM

As shown schematically in Figure 13.1, toxicants can also decrease each other's toxicity (antagonize each other). The cellular reasons are more or less the same as in the case of agonism, but the result is inhibition instead of potentiation. Antagonism may result from allosteric inhibition, in which case a chemical binds to protein changing its three-dimensional structure in such a way that the site of action of a second chemical at a different part of the protein is rendered less accessible to the second chemical (Figure 13.7). The inhibition can also result from the binding of a chemical to the same site as the second chemical, if the binding permanently changes the properties of the site of action such that the effect is reduced. The toxicity of a compound is also reduced if another toxicant speeds up its detoxification or reduces the amount of metabolites that are more toxic than the parent compound.

Similar to potentiation, a conclusion that inhibition has taken place must also be based on stringent mathematical treatment with an adequate number of observations. Since regulatory decisions are based on estimations of overall toxicities expected in an environment, it is most important for the protection of organisms that the toxicity is not underestimated. This requires that any inhibitory interactions between chemicals are on certain grounds.

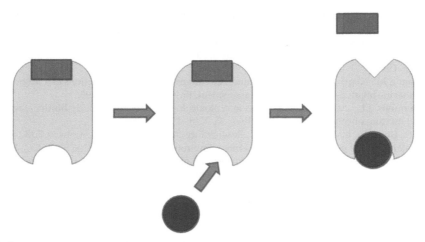

FIGURE 13.7 **Antagonistic allosterism.** The first toxicant (rectangle) causes its effect by binding to a protein and activating it. The second toxicant (circle) binds to a different site and influences the three-dimensional structure of the protein so that the first toxicant cannot bind to the protein properly and thereby cause its toxic action.

13.4 HAS POTENTIATION OR INHIBITION OF TOXICITY BY CHEMICAL INTERACTIONS BEEN DEMONSTRATED IN NATURAL ENVIRONMENTS?

While contaminants occur in various mixtures in the environment, virtually all information about the different potentiating or inhibitory effects are from well-defined, often binary, mixtures in the laboratory. Information about potentiation or inhibition of toxicant effects in the natural environment is virtually nonexistent. This is natural, since the cases where one could compare the situation of the absence of a certain toxicant with the situation of its presence in a natural setting are extremely rare. Consequently, predictions about the agonistic and antagonistic interacting effects of toxicants must be made by extrapolating laboratory findings to natural situations. Complicating factors to this are based on the fact that natural waters contain very complex mixtures of toxicants. Because of this, it is possible that all the interacting chemicals are not known. Also, it is possible that an increased number of compounds results in an interaction that is different from that of a simpler mixture that has been studied in the laboratory.

Relevant Literature and Cited References

Backhaus, T., Porsbring, T., Arrhenius, A., Brosche, S., Johansson, P., Blanck, H., 2011. Single-substance and mixture toxicity of five pharmaceuticals and personal care products to marine periphyton communities. Environ. Toxicol. Chem. 30, 2030–2040.

Bjergager, M.B., Hanson, M.L., Lissemore, L., Henriquez, N., Solomon, K.R., Cedergreen, N., 2011. Synergy in microcosms with environmentally realistic concentrations of prochloraz and esfenvalerate. Aquat. Toxicol. 101, 412–422.

Breitholtz, M., Nyholm, J.R., Karlsson, J., Andersson, P.L., 2008. Are individual NOEC levels safe for mixtures? A study on mixture toxicity of brominated flame-retardants in the copepod Nitocra spinipes. Chemosphere 72, 1242–1249.

Cedergreen, N., Kudsk, P., Mathiassen, S.K., Sorensen, H., Streibig, J.C., 2007. Reproducibility of binary-mixture toxicity studies. Environ. Toxicol. Chem. 26, 149–156.

Cedergreen, N., Kudsk, P., Mathiassen, S.K., Streibig, J.C., 2007. Combination effects of herbicides on plants and algae: Do species and test systems matter? Pest. Manag. Sci. 63, 282–295.

Celander, M.C., 2011. Cocktail effects on biomarker responses in fish. Aquat. Toxicol. 105. Suppl., 72–77.

Cleuvers, M., 2003. Aquatic ecotoxicity of pharmaceuticals including the assessment of combination effects. Toxicol. Lett. 142, 185–194.

Dondero, F., Banni, M., Negri, A., Boatti, L., Dagnino, A., Viarengo, A., 2011. Interactions of a pesticide/heavy metal mixture in marine bivalves: A transcriptomic assessment. BMC Genomics. 12, 195.

Escher, B.I., Hermens, J.L., 2002. Modes of action in ecotoxicology: Their role in body burdens, species sensitivity, QSARs, and mixture effects. Environ. Sci. Technol. 36, 4201–4217.

Gardner Jr., H.S., Brennan, L.M., Toussaint, M.W., Rosencrance, A.B., Boncavage-Hennessey, E.M., Wolfe, M.J., 1998. Environmental complex mixture toxicity assessment. Environ. Health Perspect. 106 (Suppl. 6), 1299–1305.

Ginebreda, A., Kuzmanovic, M., Guasch, H., de Alda, M.L., Lopez-Doval, J.C., Munoz, I., Ricart, M., Romani, A.M., Sabater, S., Barcelo, D., 2014. Assessment of multi-chemical pollution in aquatic ecosystems using toxic units: Compound prioritization, mixture characterization and relationships with biological descriptors. Sci. Total Environ. 468-469, 715–723.

Kreke, N., Dietrich, D.R., 2008. Physiological endpoints for potential SSRI interactions in fish. Crit. Rev. Toxicol. 38, 215–247.

Landrum, P.F., Chapman, P.M., Neff, J., Page, D.S., 2012. Evaluating the aquatic toxicity of complex organic chemical mixtures: Lessons learned from polycyclic aromatic hydrocarbon and petroleum hydrocarbon case studies. Integr. Environ. Assess. Manag. 8, 217–230.

Lang, J., Kohidai, L., 2012. Effects of the aquatic contaminant human pharmaceuticals and their mixtures on the proliferation and migratory responses of the bioindicator freshwater ciliate *Tetrahymena*. Chemosphere 89, 592–601.

LeBlanc, H.M., Culp, J.M., Baird, D.J., Alexander, A.C., Cessna, A.J., 2012. Single versus combined lethal effects of three agricultural insecticides on larvae of the freshwater insect *Chironomus dilutus*. Arch. Environ. Contam. Toxicol. 63, 378–390.

Norgaard, K.B., Cedergreen, N., 2010. Pesticide cocktails can interact synergistically on aquatic crustaceans. Environ. Sci. Pollut. Res. Int. 17, 957–967.

Schafer, R.B., Gerner, N., Kefford, B.J., Rasmussen, J.J., Beketov, M.A., de, Z.D., Liess, M., von der Ohe, P.C., 2013. How to characterize chemical exposure to predict ecologic effects on aquatic communities? Environ. Sci. Technol. 47, 7996–8004.

Schwarzenbach, R.P., Escher, B.I., Fenner, K., Hofstetter, T.B., Johnson, C.A., von, G.U., Wehrli, B., 2006. The challenge of micropollutants in aquatic systems. Science 313, 1072–1077.

Tsui, M.T., Wang, W.X., Chu, L.M., 2005. Influence of glyphosate and its formulation (Roundup) on the toxicity and bioavailability of metals to *Ceriodaphnia dubia*. Environ. Pollut. 138, 59–68.

Vellinger, C., Parant, M., Rousselle, P., Usseglio-Polatera, P., 2012. Antagonistic toxicity of arsenate and cadmium in a freshwater amphipod (*Gammarus pulex*). Ecotoxicology 21, 1817–1827.

Vighi, M., Altenburger, R., Arrhenius, A., Backhaus, T., Bodeker, W., Blanck, H., Consolaro, F., Faust, M., Finizio, A., Froehner, K., Gramatica, P., Grimme, L.H., Gronvall, F., Hamer, V., Scholze, M., Walter, H., 2003. Water quality objectives for mixtures of toxic chemicals: Problems and perspectives. Ecotoxicol. Environ. Saf. 54, 139–150.

Wacksman, M.N., Maul, J.D., Lydy, M.J., 2006. Impact of atrazine on chlorpyrifos toxicity in four aquatic vertebrates. Arch. Environ. Contam. Toxicol. 51, 681–689.

Warne, M.S., Hawker, D.W., 1995. The number of components in a mixture determines whether synergistic and antagonistic or additive toxicity predominate: The funnel hypothesis. Ecotoxicol. Environ. Saf. 31, 23–28.

Abstract

The chapter discusses the temporal effects of toxicants. Acute and chronic toxicities are ill-defined terms that should be replaced by giving the exposure times, even in the titles of studies. This information, together with information about the developmental stage of the organism, would enable accurate determination of the type of toxicity that the animal is experiencing. Normally, the toxicity experienced by organisms in the natural environment is sustained. Acute toxicity appears only in connection with spill discharges, when water purification units do not function properly, and when new chemicals are entering the environment. Most studies with acute toxicity are laboratory studies with well-defined toxicants or their mixtures. In natural environments, acute toxicity can be observed mainly in caged animals. Interpretations of immediate toxicity are complicated, as they usually contain a general stress response that may hide the responses to toxicants. Further, the use of high toxicant concentrations may result in some responses not being specific to the toxicant studied, but a more general response of a moribund organism. Some of the different responses between short- and long-term exposures of organisms to toxicants may, however, be specific to the toxicant. The short-term responses enable the organism to resist the toxicant challenge until the long-term response enables it to acclimate to the presence of the toxicant. Deciding which sublethal parameters to investigate is very important, as choosing the wrong measurement may result in an apparent absence of any exposure, even when an organism has been exposed to a toxicant.

Keywords: acute toxicity; immediate toxicity; chronic toxicity; sustained toxicity; spill discharge; stress response; stress hormones; temporal variation in toxicity; sublethal effects.

14.1 INTRODUCTION

The literature abounds with studies that have acute or chronic toxicity in the title. Consequently, it is necessary to define what acute and chronic toxicities mean. The definition

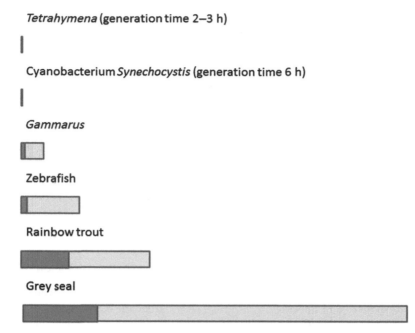

FIGURE 14.1   **The minimum generation times and life lengths of aquatic organisms differ markedly, from, say, a couple of hours to several years.** The minimal generation time is indicated in blue and the average life length in pink. Assessment of acute and chronic toxicities of different organisms must be performed with very different exposure durations to take the generation time/life length into account.

is complicated by the fact that the time to sexual maturity and life spans of organisms vary from hours to tens of years. Naturally, the length of life history of an organism must be taken into account when defining acute and chronic effects. What can constitute acute and chronic effects in different types of organisms is schematically represented in Figure 14.1. Clearly, acute and chronic effects are very different for marine mammals, long-lived fish, crustaceans with long and short generation times, unicellular algae, protozoans, and aquatic bacteria.

14.2 DIFFERENTIATING BETWEEN GENERAL STRESS RESPONSES AND SPECIFIC ACUTE RESPONSES TO POLLUTANTS

Appearance of acutely toxic effects requires that the toxicant concentration in the organism changes rapidly. In natural environments, most contamination is sustained. Rapid changes in chemical concentrations occur only with spill discharges, failures of water purification systems, or when a new contamination source discharges its effluents to a water body. Acute toxicity is thus normally restricted to either laboratory exposures with defined contaminant load or to caging exposures in which the exposed organisms are brought to the contaminated site from a clean environment. However, in caging exposures any handling stress coincides with the acute contaminant effects and, consequently, very short-term exposures cannot be assessed with reliable results about the effects of contaminants. Also, whenever acute laboratory experiments are done, an organism reacts both to the change in conditions generally

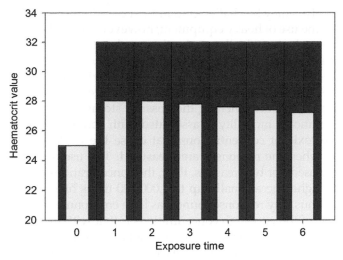

FIGURE 14.2 **The stress response hides the response to a similarly acting toxicant.** Sampling causes elevation of catecholamine concentration and consequent swelling of erythrocytes (blue; 0 indicates the hematocrit value in a blood sample taken from undisturbed fish via chronically implanted cannula, enabling the animal to remain in water throughout sampling, whereby the catecholamine concentration remains at the resting level). Blood samples are taken at time points 1–6 after netting and stunning the animal, with consequent stress-induced erythrocyte swelling. Alone, a metal exposure would cause a slight increase in cell volume (yellow bars; samples taken from undisturbed animals throughout the exposure, 0 indicates the value before the exposure; 1–6 values in metal exposed animals with 1 being the shortest exposure), but the change is hidden behind much larger stress-induced volume changes.

(general stress caused by the change in conditions) and to the toxicant added to the water specifically. In order to separate the general stress effect from the effect of the toxicant, one needs to take the same measurements (or evaluate data with the same measurements) with chemicals that should have completely different modes of action. If the responses are similar in both cases, it is likely that the responses are mainly caused by general stress, and not by the chemical that the organism is exposed to. In the case of fish, a rapid general stress reaction can be determined measuring stress hormone levels (catecholamines, corticosteroids) before and after the onset of the exposure. Catecholamine concentrations increase within a minute, and corticosteroid concentrations within a few minutes. Because of their rapid response, catecholamine concentrations usually increase as a result of the commonly used sampling procedures (netting and stunning, or anesthesia), and catecholamine-stimulated stress responses (e.g. Figure 14.2) will occur both in non-exposed and in toxicant-exposed animals. Consequently, if a toxicant effect is small and includes the same components as the catecholamine stress response, the effect may not be seen at all (see Figure 14.2). While this discussion has concentrated on fish, all animals react to rapid changes in the environment (i.e. show a general stress response) and, consequently, inappropriate handling during sampling may result in small toxicant responses not being seen. Against this background, it is surprising that quite often sampling procedures involve, for example, transport of animals to the laboratory where analyses are carried out. Instead, two alternative approaches should be taken to carrying out sampling that is minimally disturbing: First, instead of transporting the animals to the analysis laboratory, one should transport any equipment needed to the place where the animal exposures are done. This may be challenging in field conditions, e.g. obtaining electricity for centrifuges requires the use of heavy equipment; however, it must be done if truly representative samples are wanted. Second, all samples that can be stored in the frozen state should be snap-frozen in liquid nitrogen, and maintained in ultrafreezers (at −80 °C). Liquid nitrogen and ultrafreezers are advocated because the faster the freezing, the less damage occurs to the biological structures. Storage at very low temperature decreases the rate of decomposition of any biological material, thus lengthening the possible storage time.

Quite often, a significant mortality is associated with acute toxicity experiments. It is not uncommon to have toxicant concentrations that cause the death of 10, 25, and/or 50% of exposed organisms when the responses are measured. The use of such high concentrations has fortunately decreased, for two reasons. First, the concentrations causing acute lethality are usually very much higher, occasionally up to 1,000,000 times higher, than the concentrations observed in nature. Thus, any response found has little environmental relevance. The justification made for the use of the very high concentrations is that with their help the mode of action of the toxicant can be established. However, this is not certain and brings forth the other uncertainty associated with the use of concentrations associated with significant mortality: How can one be at all certain that what one measures is due to the toxicant exposure and not associated with general changes that occur in nearly dying animals? As an example, some studies have reported increases in the plasma activity of some enzymes normally occurring only in cells of fish, and associated this change with the specific toxicant that has been studied. However, necrotic cell death (and consequent release of cell contents into plasma), with the measured changes in plasma enzyme activities, will occur in all moribund fish, regardless of the reason.

14.3 TIME COURSES OF TOXICANT RESPONSES

Another significant point that must be considered is that the different responses to toxicants occur with different time courses (see Figure 12.5). Some toxicants cause responses that are initiated immediately upon contact with them. In this case, the toxicant-induced responses take place with the complement of proteins initially present in the exposed organism. In addition, the responses include transcriptional responses that can be initiated immediately at the onset of toxicant exposure. Because of the inherent time lag between transcription and translation of mRNA to protein, which varies between organisms and proteins, lasting from a couple of hours to a few days, measurable changes in protein level or activity may not occur during the short-term exposure to toxicant. This being the case, the observed changes in mRNA level, measured by quantitative polymerase chain reaction (PCR) and microarrays, are good biomarkers of exposure, but do not necessarily indicate functional responses. Thus, as discussed in Chapter 12, transcriptional changes are not biomarkers of effect. Similarly, genotoxicity, changes in the DNA structure (mutations, strand breaks, and DNA adducts) may occur immediately at the onset of toxicant exposure, but the influence of the toxicant action may first be apparent in the next generation. Consequently, choosing the measurement of sublethal effects is very important in terms of conclusions about toxicant effects. As indicated in Figure 14.3, if one measures the activity of a protein that is only affected transcriptionally by a toxicant, in a short-term acute exposure, no change in function may be observed, although events leading to decreased activity are initiated immediately upon exposure to the toxicant.

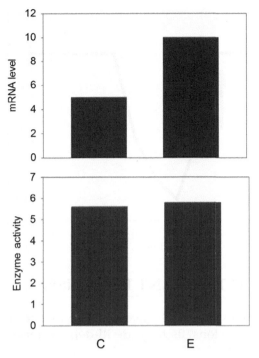

FIGURE 14.3 **A toxicant causes an immediate increase in the transcription of a gene.** The change is seen if the transcriptional response is measured by quantitative PCR (upper figure; left, mRNA level before the exposure (C); right, mRNA after 2 hours' exposure (E)). Because of the time lag between mRNA and protein production, the immediate effect of the toxicant may not be seen if protein activity is measured (lower figure; left, enzyme activity before the exposure (C); right, enzyme activity after 2 hours' exposure (E)) .

Some of the responses to toxicants are transitory. In this case, the response disappears if the contaminant exposure continues. This type of behavior is quite often seen in microarray data: short- and long-term exposures to the same toxicant are characterized by changes in the mRNA levels of quite different genes. Only a handful of the transcriptional changes observed may be the same. Transitory responses may occur:

1. If the toxicant initially increases gene transcription to increase the amount (and activity) of the protein gene product. When the transcriptional response has been associated with an adequate increase of gene product formation, it can be shut down. If the measurement of toxicant exposure is mRNA level (using, for example, quantitative PCR), the level decreases after the initial increase of transcription needed for the increased gene product formation. (To decrease completely back to the original level, it is also required that the protein stability is somewhat increased, so that the original transcriptional activity is adequate for the maintenance of an increased steady-state protein level.)
2. If the activity of a new gene product formed after exposure to toxicant is less sensitive to the toxicant than the original product (formed before toxicant exposure). In this case, as described in Figure 14.4, the toxicant initially decreases the measured activity, but upon continued exposure, new gene product is formed, and the original activity is restored.

In addition to the transitory nature of the response to a toxicant, the reason for an apparently transitory response can be the fact that, as discussed above, the immediate response to a toxicant usually involves a general response to stress, which disappears during continued exposure.

FIGURE 14.4 **A hypothetical transitory response of enzyme activity to a toxicant.** The toxicant decreases the activity of the measured enzyme (blue arrow indicates when the exposure begins). A new isoform, which is less sensitive to the toxicant, is produced as a response to the chemical exposure, and consequently the activity is restored (this is the case if the activity per unit weight of protein is restored; if the total activity is restored by an increase in the total concentration of the enzyme, then the overall gene expression has increased).

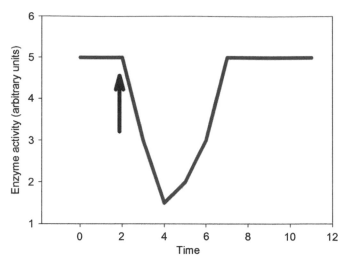

14.4 WHAT SIGNIFIES A CHRONIC TOXICANT RESPONSE?

As the exposure to a toxicant continues, it becomes chronic. Instead of using the words "acute" and "chronic" toxicities in titles and discussions, as they are ill-defined terms, a better convention would be just to state accurately the duration of toxicant exposure. This would be easy to do, as all studies give the duration of toxicant exposure. Then one would not need to define what is meant by acute and what by chronic toxicity. Notably, this direction has already been adopted to some degree, as chronic toxicity testing is often required to last more than 10% of the life length of the organism. For organisms with long lives, e.g. 10 years, this would require a more-than-1-year test. Quite often, the phrase "chronic exposure" is used when exposures with a long-living fish, such as rainbow trout, last for a month, although this period of time is much less than 10% of the life length, and even of the time from fertilization to sexual maturity. Even if a given % of the organism's life length is taken to indicate a chronic exposure, the definition does not take into account the proportion of time in the life of the organism that the development and reaching sexual maturity take. As discussed earlier (section 11.10), the toxicity of a compound may be very much affected by developmental stage, largely because active expression of genes varies markedly with developmental stage.

Most exposures in natural environments are chronic (with the notable exception of spill discharges). Acute and chronic exposures can be associated with different responses. Whereas acutely the organisms can be exposed to toxicant levels that would be lethal in chronic exposures, this cannot be done in sustained exposures (see Figure 14.5 for a scheme of different levels of toxicity). Thus, a degree of acclimation to a toxicant must have taken place, and this is probably one of the reasons for the marked difference in transcription in acute and chronic exposures. A major problem in the acute–chronic comparisons of microarray results is that in most instances one has taken only two time points, one for acute and one for chronic toxicity. It is likely that the transcriptionally induced genes are gradually replaced by others. To evaluate this possibility, a detailed time series of microarray data would be required. Obtaining

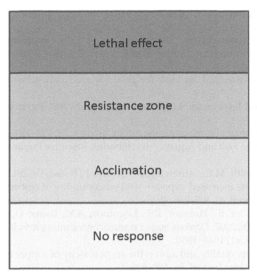

FIGURE 14.5 **The different levels of toxicity.** The darker the color, the more toxic the environment. In lethal exposure, only short-term studies can be carried out, as even adult animals die as a result of toxicant exposure. The length of time before death is observed depends on the toxicant concentration and its mode of action. Sometimes no lethality is observed during, for example, 96 hours' exposure, but all the organisms die later because of damage obtained during the exposure, even when returned to clean water. Within the resistance zone, toxicant exposures are tolerated by adult animals, but their reproduction does not succeed. Because of this, the species disappears from the community. Long-term but not multigeneration exposures are possible. In the acclimation zone, the functions of organisms are adjusted so that reproduction is possible. Here, it is possible that the initial resistance response to a toxicant is transitory and replaced by an acclimatory response, appearing later. In the no-response zone, the low-concentration toxicant exposure does not cause measurable changes in the organism.

such data carries a considerable cost, especially as one would also need to address the biological variability of the responses with adequate biological replication. Notably, even if an organism itself may be able to acclimate to the presence of a toxicant, its reproduction may be affected. Consequently, parameters related to reproduction may be most relevant, when chronic exposures are evaluated.

In addition to the responses that are sustained in the long term, some responses may only become present after a certain time of exposure. Such an observation may result if the cellular status is maintained by increased energy use. When the energy stores are depleted, any toxicant response can be manifested. Another possibility is if the toxicant response is manifested at a certain stage of development, and the experimental organisms are initially at a different developmental stage.

Although different, both immediate responses and responses appearing later may sometimes be highly specific to a particular toxicant. In this case, the immediate response aims at resisting the toxic challenge until the response appearing later enables the organism to acclimate to the presence of the toxicant.

Relevant Literature and Cited References

Ahlers, J., Riedhammer, C., Vogliano, M., Ebert, R.U., Kühne, R., Schüürmann, G., 2006. Acute to chronic ratios in aquatic toxicity: Variation across trophic levels and relationship with chemical structure. Environ. Toxicol. Chem. 25, 2937–2945.

Baun, A., Hartmann, N.B., Grieger, K., Kusk, K.O., 2008. Ecotoxicity of engineered nanoparticles to aquatic invertebrates: A brief review and recommendations for future toxicity testing. Ecotoxicology 17, 387–395.

Biales, A.D., Bencic, D.C., Flick, R.W., Lazorchak, J., Lattier, D.L., 2007. Quantification and associated variability of induced vitellogenin gene transcripts in fathead minnow (*Pimephales promelas*) by quantitative real-time polymerase chain reaction assay. Environ. Toxicol. Chem. 26, 287–296.

Biales, A.D., Kostich, M., Burgess, R.M., Ho, K.T., Bencic, D.C., Flick, R.L., Portis, L.M., Pelletier, M.C., Perron, M.M., Reiss, M., 2013. Linkage of genomic biomarkers to whole organism end points in a Toxicity Identification Evaluation (TIE). Environ. Sci. Technol. 47, 1306–1312.

Birge, W.J., Cassidy, R.A., 1983. Structure–activity relationships in aquatic toxicology. Fundam. Appl. Toxicol. 3, 359–368.

Brausch, J.M., Connors, K.A., Brooks, B.W., Rand, G.M., 2012. Human pharmaceuticals in the aquatic environment: A review of recent toxicological studies and considerations for toxicity testing. Rev. Environ. Contam. Toxicol. 218, 1–99.

Carriger, J.F., Newman, M.C., 2012. Influence diagrams as decision-making tools for pesticide risk management. Integr. Environ. Assess. Manag. 8, 339–350.

Chaisuksant, Y., Yu, Q., Connell, D.W., 1999. The internal critical level concept of nonspecific toxicity. Rev. Environ. Contam. Toxicol. 162, 1–41.

Fent, K., Weston, A.A., Caminada, D., 2006. Ecotoxicology of human pharmaceuticals. Aquat. Toxicol. 76, 122–159.

Johnson, W.W., 1980. Handbook of Acute Toxicity of Chemicals to Fish and Aquatic Invertebrates. Resource Publication 137. US Fish and Wildlife Service, Washington, DC.

Poynton, H.C., Lazorchak, J.M., Impellitteri, C.A., Blalock, B., Smith, M.E., Struewing, K., Unrine, J., Roose, D., 2013. Toxicity and transcriptomic analysis in *Hyalella azteca* suggests increased exposure and susceptibility of epibenthic organisms to zinc oxide nanoparticles. Environ. Sci. Technol. 47, 9453–9460.

Poynton, H.C., Varshavsky, J.R., Chang, B., Cavigiolio, G., Chan, S., Holman, P.S., Loguinov, A.V., Bauer, D.J., Komachi, K., Theil, E.C., Perkins, E.J., Hughes, O., Vulpe, C.D., 2007. *Daphnia magna* ecotoxicogenomics provides mechanistic insights into metal toxicity. Environ. Sci. Technol. 41, 1044–1050.

Rogevich, E.C., Hoang, T.C., Rand, G.M., 2008. The effects of water quality and age on the acute toxicity of copper to the Florida apple snail, *Pomacea paludosa*. Arch. Environ. Contam. Toxicol. 54, 690–696.

Schuler, L.J., Rand, G.M., 2008. Aquatic risk assessment of herbicides in freshwater ecosystems of South Florida. Arch. Environ. Contam. Toxicol. 54, 571–583.

Svensson, E.P. (Ed.), 2008. Aquatic Toxicology Research Focus. Nova, Hauppauge, NY.

Zhao, Y., Newman, M.C., 2006. Effects of exposure duration and recovery time during pulsed exposures. Environ. Toxicol. Chem. 25, 1298–1304.

Interactions Between Natural Environmental Factors and Toxicity

Abstract

The major abiotic and biotic factors that affect the toxicity of compounds in aquatic environments are discussed. How temperature, oxygen, salinity, pH, and other abiotic factors, as well as the biotic factors competition and predation, modify toxicological responses or are modified by them is introduced. Whenever contaminants are present in natural environments, they occur in cocktails, and the responses to them will always be affected by natural environmental variables. Quite often, environmental changes reach stressful levels and consequently the environmental stresses are confounded with toxicant stresses. Hitherto, laboratory investigations on aquatic toxicology have seldom studied toxicological responses together with normal abiotic and biotic environmental changes, but the number of studies that do take these into account is rapidly increasing. The scarce knowledge of how natural environmental changes affect toxicological responses is one factor that causes a decrease in the correspondence of biomonitoring and laboratory studies.

Temperature affects the rate of chemical reactions within the organism, whereby the detoxification and overall metabolism are temperature dependent. Temperature also influences both influx and efflux of contaminants to and from organisms. Typically, temperature changes studied experimentally are larger than naturally occurring ones, which makes their ecological relevance doubtful. Increasing temperatures, as expected to occur in the climate change scenario, are also associated with eutrophication. Variations in oxygen levels are caused by eutrophication, regardless of its cause. Interactions between contaminants and oxygen depend partially on the fact that hypoxic conditions are reducing and hyperoxic oxidizing. Changes in the redox balance influence both the stability and effects of contaminants. Many hypoxia responses are transcriptionally regulated by hypoxia-inducible factors (HIFs), which are structurally related to the aryl hydrocarbon receptor (AhR). These transcription factors may affect each other's activities, at least partly because they share the common dimerization partner aryl hydrocarbon receptor nuclear translocator (ARNT). Dimerization is required for transcriptional effects. Contaminants and oxygen transport also interact at the level of gills, as many toxicants affect gill structure and function. In addition to oxygen transfer, gills are the major site of ion and pH regulation, whereby contaminant effects on them result in interactions between contaminants and ion and pH regulation. Salinity particularly affects metal toxicity, partly because of the differences in ion uptake and efflux in marine and freshwater environments. The effects of pH on toxicity depend partly on the fact that the toxicities of the acid and base forms of weak acids are very different, with the undissociated acid form being more toxic. The interactions between intra- and interspecific competition, as well as predator–prey relationships, and toxicants are caused by several factors. First, the indirect effects of toxicants may affect the availability of hiding places. Second, the genetic variation in individual toxicant resistance is associated with changes in individual fitness of organisms, whereby intraspecific competition of organisms will be affected. Interspecific competition and predator–prey relationships are affected especially by the sensitivities of different species to contaminants affecting the overall fitness of the populations of different species.

Keywords: temperature; climate change; ocean acidification; seasonal effects; temperature acclimation; homeoviscous adaptation; oxygen; eutrophication; hypoxia; hypoxia-inducible factor (HIF); hyperoxia; gills; salinity; osmorespiratory compromise; pH; imidazole alphastat; constant relative alkalinity; UV-radiation; competition; asymmetric competition; predation; fitness; density-dependent responses.

15.1 TEMPERATURE

With regard to temperature changes in the natural environment, most laboratory studies are ecologically not very relevant, as they have studied large, rapid changes. In marine environments, such changes would only occur in shallow coastal areas and tidal pools. In the freshwater environment, rapid temperature changes could be expected only in small ponds and other small water bodies. Most temperature changes that can be expected to occur in natural environments are small, maximally a few degrees centigrade, and slow. Such changes are either not normally associated with any measurable physiological responses or the responses are hidden among the individual variations that are characteristic of the function of individuals and their populations. However, although no measurable effects may occur in the short term, species distributions can be affected in the long run, as the scope of activity of different species changes differently as a result of small temperature changes; the preferred temperature for reproduction, which affects the production of offspring, varies markedly between species (and populations within one species); etc. As an example, temperature effects on fish populations have been observed. One component of naturally occurring temperature changes is the season; because of this, the temperature responses of organisms vary seasonally. Because changes in season are most pronounced at high latitudes, where temperature changes as a result of the predicted climate change are also expected to be largest, the seasonality of responses, which probably occurs mainly in shallow-water species, must always be taken into account. In addition to the fact that both the temperature and toxicant effects can vary between different life stages, it is possible that the responses of a given stage of organisms to toxicants vary in the different seasons at the same temperature. Interacting effects of toxicants and temperature changes are most likely to occur in stenothermal organisms.

Temperature can affect the toxicity of chemicals in several major ways. First, all chemical reactions are affected by temperature change. They are speeded up by an increase in temperature, and their probability of occurrence usually increases with increasing temperature. The increase in the probability of chemical reaction usually stems from the fact that the activation energy required for the beginning of the reaction decreases (see Figure 15.1). The rate of chemical reactions in organisms normally at least doubles with every 10-degree (centigrade) increase

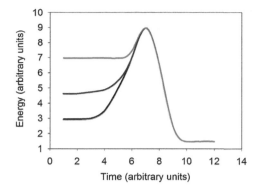

FIGURE 15.1 **Temperature effects on activation energy.** To take place, any chemical reaction must have enough energy to exceed the highest energy required for activation (highest point of the curve). The energy needed to reach this point is temperature dependent, with less energy required with an increase in temperature: green line, 30 °C; red line, 20 °C; black line, 10 °C.

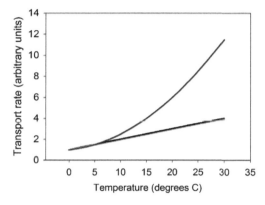

FIGURE 15.2 **Temperature and transporter function.** The figure shows a hypothetical example of how the rate of transport increases with temperature, if it occurs only be passive diffusion (black line) or if a chemical reaction in a transporter is involved (red line). Contaminants can affect the chemical reactions of the transporter, whereby the temperature effects on the transport rate will be affected.

in temperature. Thus, all the conversions of a chemical and rates of detoxification are temperature dependent. Also, all the reactions as a response to toxicant exposure will be affected by temperature change. Second, temperature has a pronounced influence on the uptake and excretion of chemicals across membranes. The influx and efflux of toxicants across membranes increase with increasing temperature. How much transport rates are affected by temperature depends on whether they occur only by passive physical diffusion or if carrier proteins with chemical reactions are involved. Purely physical passive diffusion is speeded up less with increasing temperature than transport involving chemical interaction with transporter proteins (see Figure 15.2 for a schematic example). The uptake and excretion of chemicals are of major influence to temperature-dependent changes on toxicant levels in the bodies of organisms, as the overall influx and efflux rates can have different temperature dependencies. The uptake of many chemicals is largely via the gills, and they form the majority of the uptake area of organisms. Third, the metabolism and its temperature dependence vary depending on the chemical and its metabolic pathway. An additional complicating factor in all physiological effects is temperature acclimation: it is usually considered that three weeks after the temperature change, acclimation has taken place. An important mechanism of acclimation is maintaining the fluidity of membranes (homeoviscous adaptation: Figure 15.3). Consequently, pollutants that affect membrane fluidity, such as ethanol (see section 11.7), disturb temperature acclimation. Fourth, apart from the homeothermic aquatic mammals and birds, all aquatic organisms

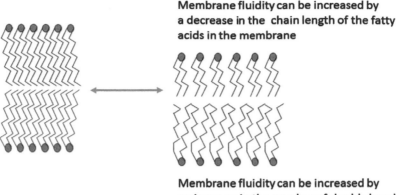

FIGURE 15.3 **Homeoviscous adaptation.** Organisms try to maintain a nearly constant fluidity of their membranes. This means that during cold acclimation, membrane fluidity increases as a result of shortening of the fatty-acid chain lengths or because of an increase in the number of double bonds in the fatty acids of membrane lipids. Any contaminants that influence the fluidity of membranes will consequently cause temperature–contaminant interactions.

FIGURE 15.4 **The scope of activity gives the difference between the maximal oxygen consumption (red line) and the resting oxygen consumption (black line) as a function of temperature.** The optimal temperature of an organism is where the difference is greatest. The organism dies when the maximal oxygen consumption falls below the resting oxygen consumption. Any contaminants affecting the oxygen consumption of organisms have larger effects when they function near the upper tolerable temperature (green circle) than when their function is close to the optimum temperature (pink circle).

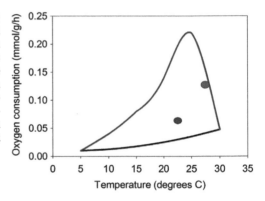

are poikilothermic. Their oxygen consumption increases with increasing temperature, and they have a certain optimal temperature range (Figure 15.4). If an organism is close to the high end of its preferred temperature range, even a small toxicant-induced disturbance in energy production can result in environmentally relevant effects. The strength of the effect of a single concentration of toxicant depends on where in the tolerated temperature range the organism resides.

15.2 OXYGEN

Both oxygen limitation (hypoxia) and hyperoxic conditions can interact with toxicological responses. Hypoxic conditions are usually reducing. In reducing conditions, for example, hydrogen sulfide, an important toxicant, is stable. Also, metals with varying valency state,

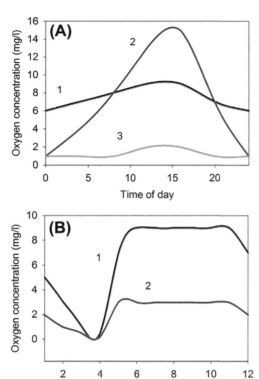

FIGURE 15.5 **Eutrophication is associated with changes in oxygen level and its variations.** (A) Daily fluctuations. (1) In a water body that is not eutrophied, daily fluctuations in oxygen concentration are small. (2) Initially, eutrophication leads to the increasing biomass of autotrophic organisms, and thus hyperoxic conditions (oxygen level is above atmospheric) prevail when photosynthesis is possible. During the night-time, all organisms consume oxygen, whereby water becomes hypoxic. The lengths of the hyperoxic and hypoxic periods depend on the lengths of the light and dark periods. (3) In highly eutrophied water, breakdown of organic matter and respiration exceed oxygen diffusion into water and its production by photosynthesis throughout the day, and water is always hypoxic. (B) The yearly fluctuations of oxygen concentration in a freezing lake. (1) During winter, the oxygen consumption by organisms and breakdown of organic matter cause hypoxic conditions, which become increasingly serious until the ice melts. Thereafter, the oxygen concentration increases and remains high until freezing again takes place. (2) In severely eutrophied water, the oxygen concentration in ice-free water does not increase to normoxic levels (air saturation more than 70%), but remains hypoxic throughout the year.

such as copper and iron, have ions with lower valency states (Cu^+ and Fe^{2+}) in hypoxic conditions. Since the toxicities of, for example, Cu^+ and Cu^{2+}, and Fe^{2+} and Fe^{3+} are different, oxygen level affects metal toxicity. The major problem with high oxygen tension is that reactive oxygen species (ROS) levels increase. Thus, high oxygen levels are associated with oxidative stress.

Hypoxic conditions occur when oxygen consumption exceeds oxygen production and diffusion. The major cause for hypoxic conditions is eutrophication. Initially, marked changes in oxygen tension occur, as in the photosynthetic zone green plants produce oxygen during the day, whereby marked hyperoxia may occur in surface waters in daytime. At night, rapid reduction of the oxygen level occurs. When eutrophication continues, the oxygen consumption of heterotrophic organisms (and oxidation of dead organic material) exceeds oxygen production by the photosynthesis of green plants, some of which also start to die, with the consequence that hypoxic conditions also prevail during the day. Figure 15.5 gives a schematic representation of the changes in oxygen level with increased eutrophication. In inland waters, two specific cases deserve special emphasis. First, in the freezing waters of temperate and boreal-zone wintertime, hypoxia occurs because ice cover prevents oxygen diffusion completely, no oxygen production by photosynthesis occurs, but all organisms respire and organic material is oxidized, albeit slowly because of the low temperature. Second, many if not most tropical inland waters are hypoxic because of the high rate of respiratory activity and oxidation of dead organic material. In addition, oxygen diffusion to the water is often limited by floating leaves or green plants. Often, only a few mm to a couple of cm of the surface water contains an appreciable amount of oxygen. In marine environments, hypoxia occurs in shallow eutrophic areas. Within the photosynthetic zone, low oxygen levels vary rhythmically (in 24-h cycles) with hyperoxic conditions. Below the photosynthetic zone, hypoxic areas occur whenever the consumption of oxygen exceeds its diffusion. This happens even when the oxygen consumption is minimal, as in the oxygen-minimum zones of oceans, usually occurring between 400- and 1000-m depth. (Below that depth, the overall oxygen consumption is so small that the oxygen level increases.)

Hyperoxic conditions occur only when photosynthetic oxygen production exceeds the overall oxygen consumption. This can be the case in eutrophic waters. Overall, oxygen as such is toxic to organisms. For this reason, effective antioxidant defenses have developed in most organisms (see section 11.3 for a detailed account of oxidative stress). Notably, changes in oxygen level are aggravated by an increase in temperature, and consequently changes in ROS levels, produced especially in hyperoxic conditions, are also implicated in responses to acute temperature increase. Thus, any toxicants affecting ROS levels will interact with both oxygen and temperature.

The interaction between oxygen and toxicant is likely to occur at the gene-regulation level. The HIF (hypoxia-inducible factor) pathway is the major regulator of hypoxia-induced gene expression (see Figure 15.6 for a schematic picture of gene regulation and the major functional targets of HIF). HIFs belong to the bHLH-PAS group of proteins, which include many environmentally regulated transcription factors, among them the aryl hydrocarbon receptor (AhR), which is the starting point of the most-studied detoxification pathway for organic chemicals with benzene rings (Chapter 9). Mammalian results indicate interaction between the HIF and AhR pathways. The described interaction of the pathways appears to be the result of the following: Both HIFαs and AhR have ARNT (aryl hydrocarbon receptor nuclear translocator) as the dimerization partner. Dimerization is required before the factors can cause transcriptional induction. The competition for the common dimerization partner results in toxicants affecting hypoxia responses and vice versa.

FIGURE 15.6 **The function and regulation of hypoxia-inducible factor (HIF).** The function of transcription factor HIF has been studied especially in mammals, and therefore knowledge of its regulation is largely based on mammalian information. It is generally accepted that HIFα, the protein moiety that confers the oxygen sensitivity to HIF, is constitutively produced. (A) In normoxia, the HIFα subunit is tagged by prolyl hydroxylase (PHD) enzyme (1) for proteasomal breakdown (2). (B) In hypoxia, PHD is inhibited, whereby HIFα is stabilized and transported to the nucleus (1). In the nucleus, HIFα forms a dimer with ARNT (2). This dimer binds to the promoter region of oxygen-sensitive genes (3), and induces their transcription (4). The induced genes are involved in, for example, blood vessel and red blood cell formation, regulation of energy metabolism, and glucose transport. Although the overall scheme of HIFα regulation appears the same in all animals, recent studies suggest that hypoxia also increases the transcription of *HIFα* in some (hypoxia-tolerant) species.

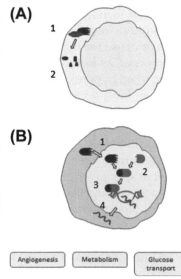

In most aquatic animals, the transfer of oxygen from the water to the organism occurs via gills. Gills maximize the gas exchange area and minimize the diffusion distance between the environment and the animal. Normally, the diffusion distance between water and blood of active teleost fish is only a couple of micrometers. Gills are also the major site of ionoregulation, whereby the responses to oxygen, salinity, and toxicants interact at the gills. Because of the importance of gills in ionoregulation, the interactions of their function with toxicants are discussed below in section 15.3, Salinity. It should also be noted that in some animals the maximal oxygen consumption at high temperatures is limited by the diffusion of oxygen through the gill epithelium. Thus, gill function also plays a role in temperature responses.

15.3 SALINITY

Another major difference between water bodies that has toxicological importance is the salinity. The major difference is between freshwater and marine environments, although a number of inland lakes are saline and can have a higher salt concentration than oceans. One reason for the variation of metal toxicity as a function of salinity is that with an increase in ion concentration, the probability of a toxic metal ion entering the site of action decreases when compared to the biologically relevant ion.

The major route of ions (as well as toxicants) to the body is via the gills. Consequently, any factors affecting the thickness of the gill epithelium or its surface area will affect both respiration and ionoregulation. When the thickness of the epithelium decreases or its area increases, the efficiency of oxygen transfer increases but the passive loss of ions also increases, requiring more energy to be put into active ion uptake in freshwater. Exactly the opposite happens if the thickness of the epithelium increases. This is often called the osmorespiratory compromise (see Figure 15.7 for the principles of gill function). The salinity of the environment affects the net direction of passive fluxes: in the marine environment, ions enter the animal passively so energy must be used to actively pump them out of the animal. Many toxicants, both metals and organic compounds, affect the diffusion distance

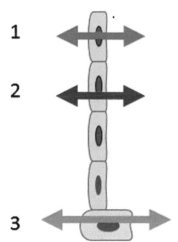

FIGURE 15.7 **Principles of gill function.** There are three major cell types in the secondary lamellae of gills: the respiratory epithelial cells, the pavement cells, and chloride cells. Gas exchange (1) and passive ion influx and efflux (2) occur mainly through respiratory epithelial cells. Contaminants can increase the thickness of the respiratory epithelium, which decreases both oxygen uptake and passive losses or gains of ions. The former effect is negative and the latter positive; this is called the osmorespiratory compromise—effects that facilitate oxygen transfer cause increase of passive ion losses or gains. Active ion uptake (freshwater) or extrusion (seawater) occurs mainly via mitochondria-rich chloride cells (3), which are situated especially at the bases of secondary lamellae.

from water to blood, which is also affected by oxygen tension. Consequently, both oxygen transfer and ionoregulation are affected. Effects on ionoregulation depend on salinity.

Salinity also affects the site of ion regulation: while gills are always important in ion regulation, the alimentary channel assumes an increasing role in marine environments. The transport sites in the gills and alimentary channel have different metal affinities with the consequence that the toxicity of metals also varies with salinity because of this.

15.4 OTHER ABIOTIC STRESSES

An important water-quality parameter that has a toxicological dimension is the acidity of water. A major aquatic environmental problem related to acidity is ocean acidification, which results from changes in the aquatic carbon dioxide–carbonate equilibrium. Also, the acidification of freshwaters has been a major question (see section 1.1). The function of virtually all enzymes, many membrane transporters, and, for example, hemoglobin is pH sensitive at physiologically relevant pH (7–8), whereby all toxicant effects on their function will also be pH sensitive. The reason why pH changes are very important in the regulation of protein function stems from the fact that the charge of their amino acids is the major determinant of their three-dimensional structure and, consequently, activity. Of the amino acids, histidine is particularly important, since its protonation (histidine-imidazole charge) changes in the physiological pH range. In fact, the so-called imidazole alphastat (or constant relative alkalinity) hypothesis, based on the maintenance of histidine-imidazole charge, was formulated to explain the acid–base regulation of animals. In addition to histidine imidazole, the charge of sulfhydryl groups (–SH) and primary amino groups (–NH2) changes in the physiological pH range, 7–8. Notably, pH is a highly temperature-sensitive parameter. Neutral water has a pH of 7 at 25 °C. As shown in Figure 15.8, the pH of neutral water decreases with increasing temperature, and increases with decreasing temperature.

In aquatic animals, pH regulation is mainly carried out by the gills. The ion exchangers regulating the ion balance also control the acid–base balance. Consequently, ion and acid–base regulation are disturbed by the same environmental toxicants, and environmental salinity

FIGURE 15.8 **The dependence of neutral water pH on temperature.**

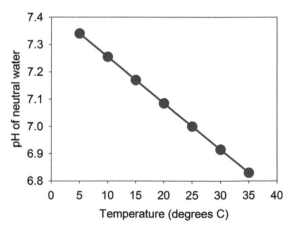

and pH influence toxicant effects on both ion and acid–base regulation. An additional factor with regard to water pH/toxicant response interactions is that the toxicity of all weak acids and bases varies markedly with their dissociation status: usually the undissociated form is more toxic than the charged form. The uptake of chemicals in organisms depends mainly on their lipophilicity, which is much higher for the acid (undissociated) than base (dissociated) forms of weak acids/bases (see Chapter 7). Consequently, they also distribute more readily to membranes than cytoplasm and can, for example, exert membrane-dependent toxicity.

An important abiotic factor modifying the toxicity of chemicals is ultraviolet (UV) radiation. While the effect is restricted to surface waters because of the short penetration distance of UV radiation in water, several chemicals undergo degradation in sunlight. Some of the daughter products are more toxic than the parent compound. A particular group of compounds to which attention is directed is sunscreens, which attain high concentrations in the coastal waters of holiday resorts, and are often subject to photolysis.

15.5 COMPETITION AND PREDATION

In intraspecific competition, individuals of the same species compete against each other for food, mates, and shelter. The population structure and size is determined by this. The population density that can be attained depends on the so-called carrying capacity of the given environment for the species. Important determinants of the carrying capacity are the availabilities of food and shelter, which can both be affected by contamination. Often, the intraspecific competition is affected by density: it may not occur at all at low population density but can be pronounced when the carrying capacity of an environment for the species is approached. The effects of toxicants on intraspecific competition depend largely on whether the chemicals affect the individuals differently. The differences in the genetic composition between individuals may be behind different responses of individuals to toxicant responses. It appears that in many cases the success of individuals in a contaminated environment depends mainly on certain gene products being compatible with life in the contaminated environment. In this case, the fitness of offspring carrying the same trait is increased as compared with the offspring of other individuals. As illustrated in Figure 15.9, the end result may be that the trait enabling the organism to tolerate the contaminated environment is rapidly enriched in a population, with no requirement for the occurrence of a rare beneficial mutation. Thus, carrying a trait that enables the organism to tolerate a contaminated environment gives a significant competitive edge. This type of genetic response is especially important for species with a long generation interval, as new mutations are not required. Species with short generation times can gain advantage from rare beneficial mutations. Their occurrence has been described for aquatic bacteria and unicellular algae. Most of the studies that have shown a genetic adaptation to contaminated conditions have not differentiated between a change in the mutation rate and increased survival of organisms with a trait that was already present.

Interspecific competition means that individuals of one species suffer a reduction of fecundity, growth, and/or survival because of the resource use by another species. When different species have different susceptibilities to a contaminant, interspecific competition is also affected. Furthermore, the competitive edge of a species may depend on temperature or other abiotic factors. These abiotic factors may also affect the toxicant effects on

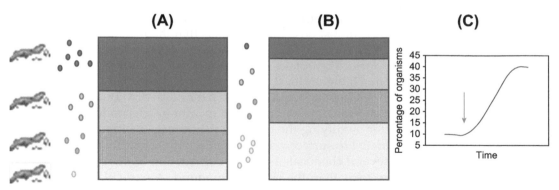

FIGURE 15.9   **Enrichment of a previously existing trait in a population enabling tolerance of a contaminated environment.** Four types of animals exist in a population. Of these, the blue type has the highest efficiency of reproduction in the absence of contaminant (A), constituting 40% of the population. The animals of the green type make up 30%, those of pink type 20% and yellow type 10%. When exposed to a toxicant (B), the efficiency of reproduction between animal types changes markedly so that the yellow type has highest reproductive efficiency, now making up 40% of the population, the green and pink types each 25%, and the blue type 10%. Thus, the toxicant increases the competitive edge of the yellow type markedly. (C) The proportion of yellow-type organisms in the population. In the absence of the toxicant, the population is stable and the yellow type makes up 10%. At the arrow the toxicant is added, and the proportion of the yellow type increases until the population again becomes stable, at which point the yellow type makes up 40% of the population.

the species. Consequently, the result is a complex interaction of different biotic, abiotic, and contaminant effects. Competition may be such that one species exploits the environment in a way that will affect other species. An example of such competition is where an aquatic herbivore affects the vegetation in such a way that hiding places for another species decrease in number, whereby that species will be more vulnerable to predation. This is a clear example of indirect, asymmetric competition. The species do not compete for the same resource, but utilization (exploitation) of a resource by one species leads to an effect on another. Such competition is highly asymmetric, as the herbivore may not be affected at all, while the survival of the other species is significantly affected. It is important to note that studying the effects of contaminants on competition necessarily requires multispecies experimentation. Further, as the responses to a contaminant may be dependent on temperature and other abiotic factors such as day length, carrying out a mesocosm experiment in summer can conceivably give a different result from the same kind of experiment done in winter. Another point is that one may not know all the components in a competition, and therefore even the best mesocosm experiment is sometimes a poor substitute for the real environment. Competitive effects may be studied within one generation. This, however, leaves any transgenerational interactive effects of competition and contaminants unexplored. On the other hand, multigenerational studies can only be done with species having short generation times, such as prokaryotes, unicellular algae, and protozoans. This necessarily generates bias for the types of responses that are observed, and if the transgenerational responses of species with long generation times are different from those of organisms with short generation times, leads to inaccurate conclusions.

Predation is the consumption of one organism (the prey) by another (the predator). Predators can be classified as:

1. True predators. These usually kill their prey, and consume many prey organisms during their lifetime. Carnivores, animal-eating organisms, are typical true predators.
2. Grazers. These consume only part of their prey at one go, but consume (parts from) many prey organisms during their lifetime. Herbivores, plant-eating animals, are typical examples of grazers.
3. Parasites. These consume parts of their prey (host), and are usually restricted to one or a few hosts during their lifetime. Because they are restricted to one host, their effect must be nonlethal in the short term. They are also characterized by complex life cycles including primary and secondary hosts.
4. Decomposers/detritivores. These are species that use dead organisms to obtain energy and material for anabolic reactions.

The simplest way by which predation can be influenced by toxicants is that the sensitivities of predator and prey to the compound are different. As a consequence, and depending on the function that is affected, either the population density of the predator or its preying efficiency is affected. The latter are commonly studied as behavioral endpoints (e.g. time required to catch prey, attempts required before prey is caught). Many prey organisms, especially plants, have "protective compounds" released from the cells when the cell is broken. The release of these, often phenolic compounds in plants, can be affected by toxicants, affecting their preference as the prey. As with the studies on competition, predator–prey relationships can naturally be studied only with multispecies study designs, and one also needs to remember that adding complexity to the design adds complexity to the analysis, with the consequence that unequivocal conclusions may not be possible. Further, as with competition, predation effects are density dependent, with both the predator and prey densities having an effect.

Relevant Literature and Cited References

Agbo, S.O., Keinanen, M., Keski-Saari, S., Lemmetyinen, J., Akkanen, J., Leppanen, M.T., Mayer, P., Kukkonen, J.V., 2013. Changes in *Lumbriculus variegatus* metabolites under hypoxic exposure to benzo(a)pyrene, chlorpyrifos and pentachlorophenol: Consequences on biotransformation. Chemosphere 93, 302–310.

Begon, M., Townsend, C.R., Harper, J.L., 2006. Ecology: From Individuals to Ecosystems, fourth ed. Blackwell, Oxford, UK.

Byrne, M., 2012. Global change ecotoxicology: Identification of early life history bottlenecks in marine invertebrates, variable species responses and variable experimental approaches. Mar. Environ. Res. 76, 3–15.

Byrne, M., Przeslawski, R., 2013. Multistressor impacts of warming and acidification of the ocean on marine invertebrates' life histories. Integr. Comp. Biol. 53, 582–596.

Byrne, M., Soars, N., Selvakumaraswamy, P., Dworjanyn, S.A., Davis, A.R., 2010. Sea urchin fertilization in a warm, acidified and high pCO2 ocean across a range of sperm densities. Mar. Environ. Res. 69, 234–239.

Cairns Jr., J., Heath, A.G., Parker, B.C., 1975. Temperature influence on chemical toxicity to aquatic organisms. J. Water Pollut. Control Fed. 47, 267–280.

Eisler, R., 2012. Oceanic Acidification: A Comprehensive Overview. Science. St. Helier, Jersey, UK.

Feidantsis, K., Antonopoulou, E., Lazou, A., Portner, H.O., Michaelidis, B., 2013. Seasonal variations of cellular stress response of the gilthead sea bream (*Sparus aurata*). J. Comp. Physiol. B.—Biochem. Syst. Environ. Physiol. 183, 625–639.

Fenchel, T., Finlay, B., 2008. Oxygen and the spatial structure of microbial communities. Biol Rev. Camb. Philos. Soc. 83, 553–569.

Heise, K., Puntarulo, S., Nikinmaa, M., Abele, D., Portner, H.O., 2006. Oxidative stress during stressful heat exposure and recovery in the North Sea eelpout *Zoarces viviparus* L. J. Exp. Biol. 209, 353–363.

Ivanina, A.V., Kurochkin, I.O., Leamy, L., Sokolova, I.M., 2012. Effects of temperature and cadmium exposure on the mitochondria of oysters (*Crassostrea virginica*) exposed to hypoxia and subsequent reoxygenation. J. Exp. Biol. 215, 3142–3154.

Kreitsberg, R., Barsiene, J., Freiberg, R., Andreikenaite, L., Tammaru, T., Rumvolt, K., Tuvikene, A., 2013. Biomarkers of effects of hypoxia and oil-shale-contaminated sediments in laboratory-exposed gibel carp (*Carassius auratus gibelio*). Ecotoxicol. Environ. Saf. 98, 227–235.

Lemly, A.D., 1996. Winter stress syndrome: An important consideration for hazard assessment of aquatic pollutants. Ecotoxicol. Environ. Saf. 34, 223–227.

Lockhart, W.L., 1995. Implications of chemical contaminants for aquatic animals in the Canadian arctic: Some review comments. Sci. Total Environ. 160-161, 631–641.

Lyu, K., Cao, H., Chen, R., Wang, Q., Yang, Z., 2013. Combined effects of hypoxia and ammonia to *Daphnia similis* estimated with life-history traits. Environ. Sci. Pollut. Res. Int. 20, 5379–5387.

Mustafa, S.A., Davies, S.J., Jha, A.N., 2012. Determination of hypoxia and dietary copper mediated sub-lethal toxicity in carp, *Cyprinus carpio*, at different levels of biological organisation. Chemosphere 87, 413–422.

Negreiros, L.A., Silva, B.F., Paulino, M.G., Fernandes, M.N., Chippari-Gomes, A.R., 2011. Effects of hypoxia and petroleum on the genotoxic and morphological parameters of *Hippocampus reidi*. Comp. Biochem. Physiol. C. Toxicol. Pharmacol. 153, 408–411.

Nikinmaa, M., 2013. Climate change and ocean acidification—interactions with aquatic toxicology. Aquat. Toxicol. 126, 365–372.

Noyes, P.D., McElwee, M.K., Miller, H.D., Clark, B.W., van Tiem, L.A., Walcott, K.C., Erwin, K.N., Levin, E.D., 2009. The toxicology of climate change: Environmental contaminants in a warming world. Environ. Int. 35, 971–986.

Portner, H.O., 2010. Oxygen- and capacity-limitation of thermal tolerance: A matrix for integrating climate-related stressor effects in marine ecosystems. J. Exp. Biol. 213, 881–893.

Portner, H.O., Peck, M.A., 2010. Climate change effects on fishes and fisheries: Towards a cause-and-effect understanding. J. Fish Biol. 77, 1745–1779.

Randall, D.J., Tsui, T.K., 2002. Ammonia toxicity in fish. Mar. Pollut. Bull 45, 17–23.

Richards, J.F., Farrell, A.P., Brauner, C.J. (Eds.), 2009. Hypoxia. Fish Physiology, vol. 27. Elsevier, San Diego, CA.

Saarikoski, J., Lindström, R., Tyynelä, M., Viluksela, M., 1986. Factors affecting the absorption of phenolics and carboxylic acids in the guppy (*Poecilia reticulata*). Ecotoxicol. Environ. Saf. 11, 158–173.

Saarikoski, J., Viluksela, M., 1981. Influence of pH on toxicity of substituted phenols to fish. Arch. Environm. Contam. Toxicol. 10, 747–753.

Schiedek, D., Sundelin, B., Readman, J.W., Macdonald, R.W., 2007. Interactions between climate change and contaminants. Mar. Pollut. Bull 54, 1845–1856.

Sharma, V.K., Siskova, K.M., Zboril, R., Gardea-Torresdey, J.L., 2014. Organic-coated silver nanoparticles in biological and environmental conditions: Fate, stability and toxicity. Adv. Colloid Interface Sci. 204, 15–34.

Valavanidis, A., Vlahogianni, T., Dassenakis, M., Scoullos, M., 2006. Molecular biomarkers of oxidative stress in aquatic organisms in relation to toxic environmental pollutants. Ecotoxicol. Environ. Saf. 64, 178–189.

Van der Geest, H.G., Soppe, W.J., Greve, G.D., Kroon, A., Kraak, M.H., 2002. Combined effects of lowered oxygen and toxicants (copper and diazinon) on the mayfly *Ephoron virgo*. Environ. Toxicol. Chem. 21, 431–436.

Effects of Chemicals on Aquatic Populations

Abstract

Toxicant effects on individuals are ecologically meaningful only if the populations are also affected. Conclusions about effects of toxicants are complicated if immigration is possible, which can result in a lack of population-level changes even if the individuals are negatively affected by toxicants and their mortality increases. The simplest metrics that can be applied to the population effects of toxicants are the population size and growth rate. When considering the effects of contaminants on populations, the life history properties, such as time to sexual maturation, total offspring production, and average life length, must be taken into account whenever comparisons between species are made. A problem with translating results from laboratory experiments to predictions about the behavior of wild populations is that most laboratory experiments do not consider density-dependent phenomena that may be important for the responses of wild populations to toxicants. Genetic variability is an important component of populations, and often affected by contaminants. The first generation exposed to toxicants can tolerate them if their capacity to acclimate (phenotypic plasticity) is adequate. Further survival of the population is best if it is highly heterozygous. The development of toxicant tolerance is a good example of directional natural selection, and is fastest if the trait conferring tolerance is common even without the exposure.

Keywords: Euler–Lotka equation; Hardy–Weinberg principle; reproductive efficiency; immigration; metapopulation; epidemiology; incidence; prevalence; relative risk; correlation; density-dependent effects; phenotypic plasticity; genetic bottleneck; heterozygosity; heritability; toxicant tolerance; toxicant resistance.

An Introduction to Aquatic Toxicology
http://dx.doi.org/10.1016/B978-0-12-411574-3.00016-5

16.1 INTRODUCTION

Although much of the work done in aquatic toxicology is done in the laboratory and with single species, and often subcellular or suborganismic effects are determined, it must be remembered that an effect is only relevant ecologically if contamination affects a population of organisms. A population comprises a number of individuals of a species in a defined space. Also, as discussed previously, contaminants do not occur singly in natural environments, but as complex cocktails, and always interact with natural environmental variations. Thus, while the most exact knowledge of toxicant effects can be obtained at the molecular level, this is least relevant for environmental decisions unless the results can be anchored to ecological consequences (Figure 16.1). A further important point to take into account is that an ecological effect can only take place if some molecular function is affected. Thus, aquatic toxicology is truly integrative. While it must be accepted that every single scientist can only work at some portion of the whole ecotoxicological landscape, it is equally important to accept that no single traditional discipline is adequate for solving how contaminants affect populations and ecosystems and what environmental decisions must be taken.

When effects of contamination on a population are considered, it is important to note that if the size and reproduction of the population are not affected, there is no contaminant effect. The reproductive efficiency must be included when natural populations are considered, since the size of a population can remain constant even if the contaminant affects the population, if the decreased reproduction or increased mortality is compensated for by immigration (see Figure 16.2 for a general scheme of how populations can be affected by toxicants). The efficiency of immigration in replacing lost individuals depends on how migrant an organism is. Thus, it is more likely for pelagic fish than sessile mussels. Now, if only population size were taken to indicate the toxicant effect, an erroneous conclusion could be reached, as with migrant species immigration could replace any lost organisms, whereby toxicity to a sessile animal would be considered higher than to a migrant species even when they are actually the same. Immigration and emigration events can also have an effect on the estimated overall exposure. If organisms are living in a patchy population, and the different patches have different exposure histories, the overall movement of organisms between patches will affect both the apparent exposure and the apparent effects. The effect decreases with a decrease of movement between patches. When

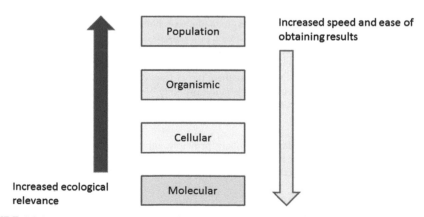

FIGURE 16.1   The study types in aquatic toxicology, from the molecular to the population level.

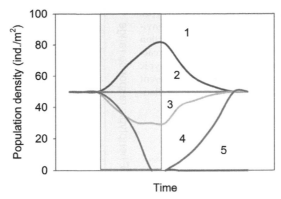

FIGURE 16.2 **Possible effects of contamination on population density.** The colored area indicates the period of contaminant exposure. (1) Contaminant exposure increases the population density of a species. The density decreases back to the original level after the exposure is discontinued. Originally the population is stable. With contaminant exposure, the fitness (efficiency of reproduction) of the studied species increases, whereby its density increases to a higher steady-state value. (2) The contaminant exposure does not affect the population density. This can be caused either by the contaminant not affecting the fitness of the species or by the reduced recruitment being replaced by increased immigration. (3) Contaminant exposure causes a decline in the population of species with a recovery upon cessation of the exposure. (4) Exposure to the contaminant causes a local extinction of the species with recovery of population through immigration after the discontinuation of the exposure. (5) The extinction caused by contamination persists in the absence of immigration.

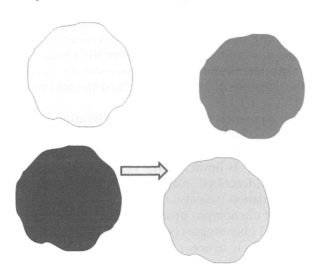

FIGURE 16.3 **A population consists of four metapopulations, all of which have different contaminant exposures (the darker the color, the more exposure).** An individual that has experienced the most intensive exposure is immigrating to an area with a metapopulation that has experienced less contamination. If a response is measured from such an individual and related to where it is presently living, a faulty conclusion about exposure and response is reached.

organisms are most of the time living as discrete populations, but immigration from one population to another is possible and occasionally occurs, one is talking of metapopulations. The different metapopulations may have drastically different exposure histories (see Figure 16.3). Thus, migration from one metapopulation to another also affects the overall exposure experienced by the organisms.

If immigration or emigration is not possible in the short term, the population of a species may be affected as described schematically in Figure 16.2. To be able to evaluate which

FIGURE 16.4 **Phenotypic plasticity.** Three genotypes are given as differently colored lines. The plausible environmental changes are shown with pink. The proportion of the genotype lines in the pink area gives the reaction norm, which translates to the possible phenotypic responses, given in green, yellow, and blue. The different genotypes are characterized by different ranges of plausible phenotypes.

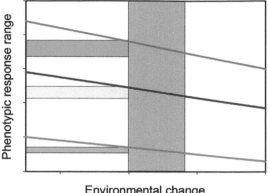

alternative results from contaminant exposure, it is imperative that population size/density can be determined. The most accurate, but in most cases impossible, alternative is to count all individuals in the population of a given space. Consequently, the population density is normally determined from smaller samples. For fish, population estimation usually involves initial catching of fish, often with gill nets with varying mesh size. However, electrofishing, trawling, etc., can also be used. The fish are marked and released, and the population size can be estimated from the proportion of marked fish appearing in a new catch (mark–recapture method). In addition to the population size, the mark–recapture method can be used to estimate the age distribution of fish. To obtain an estimation of population densities of plankton species, several samples need to be taken from different parts of the water column (different depths, different locations) so that the sampling takes into account the different environments for any species. Similarly, benthic populations need to be sampled at different locations to cover the densities of organisms at the different bottom types.

The effects of contaminants on populations depend on how plastic the individual phenotypes in the population are. If the individuals of a population show reversible acclimation to toxicant exposure, the overall population response remains small. If, on the other hand, the phenotypic plasticity of individuals in a population is limited, large effects at the population level can be observed. Phenotypic plasticity (described in Figure 16.4) in a population depends on the genetic composition of the population (see also section 16.4). The relationship between environmental change and range of phenotypes produced for a genetic type of organism (the same genotype can produce several phenotypes) is determined by its reaction norm, and genetically distinct reaction norms are likely to occur.

16.2 EPIDEMIOLOGY

Although epidemiology was originally developed to understand how diseases are spread and affect (human) populations, the treatments used are for the most part usable when considering population-level effects of toxicants. An epidemiological approach has been used successfully, for example, when the incidence of tumors in flatfish inhabiting coastal areas with heavy polyaromatic hydrocarbon pollution has been looked at.

The likelihood of an individual of a population showing an effect if exposed to a toxicant once (but with varying time) is given by the incidence rate (I), where:

$$I = \frac{N}{T},$$

(16.1)

where N is the number of individuals showing an effect and T is the total time that the population has been exposed (in human epidemiology, the result is e.g. 10 cases of disease when the total exposure time to the disease-causing agent is 100 person years). The prevalence (P) is then the incidence rate multiplied by the time that individuals were actually at risk (t):

$$P = I \times t.$$

(16.2)

Thus the incidence rate gives the risk, while the prevalence actually shows the likelihood that the risk is realized in the scenario taking place. Comparing the risk (incidence rate) between control (I_c) and exposed (I_e) populations gives the relative risk (RR):

$$RR = \frac{I_e}{I_c}.$$

(16.3)

In epidemiology, one is often talking about the association of a disease (in aquatic toxicology, an observed effect) with a risk factor. Usually the associations are correlational. Thus, one cannot be sure that an association actually gives a cause–effect relationship or merely indicates that both associated phenomena depend on a third factor (see Figure 16.5 for a schematic representation of reasons behind correlations). The strength of the association is

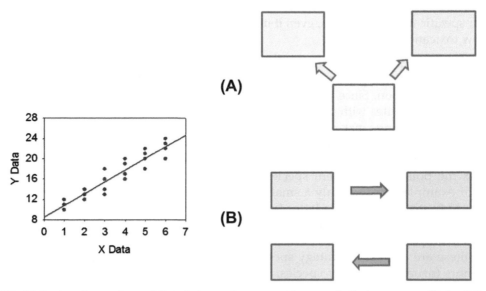

FIGURE 16.5 **An observed correlation between two parameters can indicate a cause–effect relationship between the two, but equally well there does not need to be a relationship between the two.** (A) Both (blue, pink) may depend on additional factor (green). (B) Also, while blue can be caused by lilac, it is equally possible that lilac causes blue (reverse causation).

increased if a dose–response relationship between the risk factor and disease state can be obtained. Similarly, if manipulations indicate a mechanistic (cause–effect) link between a risk factor and an effect that is seen in individuals, the strength of the association is markedly increased.

16.3 DEMOGRAPHIC EFFECTS

When one is considering effects of toxicants on populations, different aspects of demographic characteristics can be important end points for any characterization of chemical effects. The simplest demographic indices are the total population size and the total growth rate of the population, given by the Euler–Lotka equation:

$$1 = \sum_{a=1}^{\omega} \lambda^{-a} l(a) b(a)$$

(16.4)

where λ is the discrete growth rate, $l(a)$ is the fraction of individuals surviving to age a, and $b(a)$ is the number of individuals born at time a. In addition, at least the following characteristics can be important for indicating contaminant effects:

1. Time to sexual maturation, i.e. the time from birth to the first production of offspring.
2. The age-specific birth rate. It is possible that even when the total production of offspring is not affected, the birth rate at a given age changes.
3. Total offspring production.
4. Mean and median life length.
5. Age-specific mortality, which, even if mean life length is not affected by contaminant, can show toxicant effects.

In estimating demographic effects, one should be able to characterize them before contamination and after exposure to be able to show unequivocally that observed findings are due to contamination. Since this is not possible in most instances, data on both contaminated and control sites with similar water qualities should be obtained. Predictions based on laboratory findings may not be at all accurate for natural populations for a couple of reasons. First, it is usually thought that if similar effects on mortality are observed on two species, then the overall population effects are similar. However, this contention fails to address the point that different species may have markedly different life histories: one may, for example, produce only a small number of offspring (K strategy) and the other a multitude (r strategy). For the latter type of organism, it does not really matter if it dies after successful reproduction; for the former, the survival of the parent is important. As another example, if one species has a very long period before the first reproduction takes place (these are usually K-strategy species, which are normally also long living), another very short (usually r-strategy species, which normally have also short generation time), the production of offspring will be very differently affected by mortality in the two species, if the life history characteristics are not taken into account (by, for example, giving the mortality before sexual maturation as the end point that is compared in the two species). Second, laboratory studies seldom address density-dependent phenomena; normally the food availability (energy availability) or availability of space is not a limiting factor. (Note, however, that in many cases, laboratory experiments with fish and many invertebrates are done with unfed animals.) In natural environments, both of these may limit the success of a population, and it is also quite common that reproductive efficiency is influenced by population density. Notably, the models used in aquatic toxicology usually do not address the density dependency of simulated phenomena. Thus, one is left with a big uncertainty: whether the observations made in unlimiting conditions of food and space reflect the situation of natural populations, where both are likely to be limiting factors at some time during the life span of the organisms in a population. For example, toxicities of compounds may be markedly different between a starving, a reasonably fed, and a fat population. Third, the population responses may be different in growing, steady-state, and declining populations.

16.4 POPULATION GENETICS

Any population is composed of individuals with different genetic structure. Contamination affects the genetic composition as described in Figure 16.6. When one is protecting a population, one of the most important considerations is protecting genetic variability. It is commonly accepted that contamination affects the genetic variability within a population. Large equilibrium populations with negligible mutation and migration rates follow the Hardy–Weinberg principle, stating that the frequency of genotypes is constant.

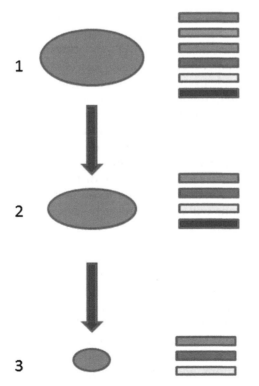

FIGURE 16.6 **Contamination effects on genetic variability.** (1) In the absence of contaminant, the population is genetically diverse (indicated by six colored bars). (2) Upon exposure to toxicant, the individuals that are most susceptible to the contaminant die off, whereby the population size decreases slightly (the size of the ellipse denoting population size is somewhat reduced) and the genetic diversity is decreased (indicated by a decrease in the number of bars from six to four). (3) The efficiency of reproduction is reduced, whereby the population size is reduced (smaller size of the ellipse), and the genetic diversity is further reduced (only three bars remaining).

If contamination affects the genetic composition of a population, the statistical significance of the change can be calculated with the χ^2 test. Usually, results are available for loci with two or three alleles. In the case of three alleles, the equation is the following:

$$\chi^2 = \sum_{i=1}^{6} (O_i - E_i)^2 / E_i.$$

(16.5)

The degrees of freedom for χ^2 equal the number of possible genotypes (six in the case of three alleles) minus the number of alleles (three). O_i = observed allele frequency and E_i = expected allele frequency for allele i.

If there is a large genetic variability, i.e. the population is highly heterozygous, it is likely that some individuals of the population will have genetic properties that allow their survival/success in a contaminated environment. It is generally accepted that the heterozygosity decreases with contaminant load. Typically, the contamination decreases the effective population size (i.e. the population that carries genetic properties to the next generation). Occasionally, contamination reduces the population so drastically that significant genetic bottlenecks are formed. The heterozygosity of the population surviving the exposure to a contaminant may be so much reduced that the population cannot survive another type of stressor (or contaminant) that the parent population, which has not experienced the contaminant-induced reduction in the effective population size (and consequent decrease in heterozygosity), can easily survive.

In the first generation responding to a contaminant, the survival of a population is dependent on acclimation via phenotypic responses, but transgenerational survival requires that the responses allowing survival are heritable. In the narrow sense, heritability (which is probably the most relevant measure in the context of contaminant-exposed populations, since it does not require that the trait conferring survival is normally distributed) can be estimated from linear regression of the measured trait in the offspring and in a parent. The heritability can also be estimated with the help of family trees and measurements made from them in different generations. In the best case, this enables evaluation of how environmental changes have affected the genetics of populations. The major weakness in this regard is that good, long-term family trees are seldom available for wild organisms. It is important to note that transgenerational influences do not necessarily require a change in DNA structure: if changes in histone modification/DNA methylation persist over successive generations, a transgenerational effect is achieved epigenetically.

Changes in the genetic constituents in a population can occur via genetic drift or natural selection. As the name implies, genetic drift is not directional, and is not a result of a genetic response to a stimulus. However, two distinct populations will differ from each other because of genetic drift, because it is highly unlikely that the same genetic changes would occur by chance. If populations are small (and consequently have reduced heterozygosity), the likelihood of genetic-drift-induced difference between the populations is increased, as the likelihood of both populations having the same alleles originally is decreased. Because of the genetic differences in the populations, they can also respond differently to contaminants. If they do, the difference has come about by chance. Directional genetic responses to contaminants require that natural selection occurs. Selection can be directional, stabilizing, or disruptive (see Figure 16.7).

Toxicant resistance (or toxicant tolerance; the two phrases are taken here to be synonymous) is mainly the consequence of directional selection. It is developed mostly with existing genetic components; only in rare cases are mutations generating new gene products

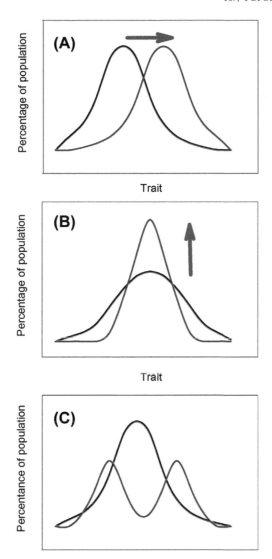

FIGURE 16.7 **Selection types.** Black lines indicate the properties of a trait before selection and red lines after selection. (A) Directional selection. The arrow indicates the direction of change of the trait. (B) Stabilizing selection. The arrow indicates that the variance of the trait decreases. (C) Disruptive selection. The trait is divided into two peaks.

and responses involved. Mutations increasing toxicant tolerance occur mainly in organisms with a short generation time. The acquisition of tolerance is affected at least by the following factors:

1. If the genetic property that is modified in acquiring toxicant tolerance is common, tolerance is developed fast.
2. If the development of tolerance depends only on one gene, it is developed faster than if the development requires several genes.

FIGURE 16.8 **Development of toxicant tolerance.** (A) A polygenic trait. Tolerance is greatest on the right. (1) Before toxicant exposure, the population is least tolerant. With increasing time of exposure (2, 3), the toxicant tolerance is increased via directional selection. (B) A monogenic trait. The trait indicated with a green bar is the most tolerant to the toxicant. Its proportion increases with exposure time: (1) the proportion of traits before toxicant exposure; (2, 3) toxicant-exposed organisms (with longer exposure of 3 than 2).

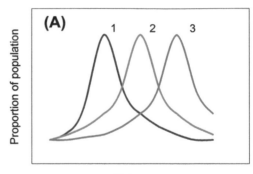

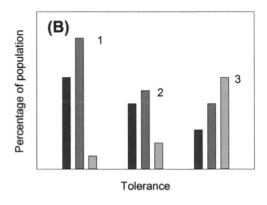

3. If the genetic component that is required to change is associated with a factor that decreases tolerance, the overall development of tolerance is slowed down.
4. The development of tolerance is speeded up by increasing difference in the fitness of tolerant and non-tolerant specimens.
5. The development of tolerance is speeded up by increased number of offspring and by shortened generation time of the organism.

Two life-history effects will slow down the development of tolerance: first, if the environment has places where the organism can escape the exposure to contaminant, and second, if significant immigration of non-tolerant individuals occurs. The first alternative is true for any sites upstream to toxicant efflux. The development of toxicant tolerance is illustrated in Figure 16.8.

Relevant Literature and Cited References

Bontje, D., Kooi, B.W., Liebig, M., Kooijman, S.A., 2009. Modelling long-term ecotoxicological effects on an algal population under dynamic nutrient stress. Water Res. 43, 3292–3300.

Brausch, J.M., Salice, C.J., 2011. Effects of an environmentally realistic pesticide mixture on *Daphnia magna* exposed for two generations. Arch. Environ. Contam. Toxicol. 61, 272–279.

Dahiya, R.C., 1980. Estimating the population sizes of different types of organisms in a plankton sample. Biometrics 36, 437–446.

Foit, K., Kaske, O., Wahrendorf, D.S., Duquesne, S., Liess, M., 2012. Automated Nanocosm test system to assess the effects of stressors on two interacting populations. Aquat. Toxicol. 109, 243–249.

Forbes, V.E. (Ed.), 1999. Genetics and Ecotoxicology. Taylor and Francis, Philadelphia, PA.

Guttman, S.I., 1994. Population genetic structure and ecotoxicology. Environ. Health Perspect. 102 (Suppl. 12), 97–100.

Kohler, H.R., Triebskorn, R., 2013. Wildlife ecotoxicology of pesticides: Can we track effects to the population level and beyond? Science 341, 759–765.

Martinez-Jeronimo, F., Arzate-Cardenas, M., Ortiz-Butron, R., 2013. Linking sub-individual and population level toxicity effects in *Daphnia schoedleri* (Cladocera: Anomopoda) exposed to sublethal concentrations of the pesticide alpha-cypermethrin. Ecotoxicology 22, 985–995.

Messiaen, M., De Schamphelaere, K.A., Muyssen, B.T., Janssen, C.R., 2010. The micro-evolutionary potential of *Daphnia magna* population exposed to temperature and cadmium stress. Ecotoxicol. Environ. Saf. 73, 1114–1122.

Messiaen, M., Janssen, C.R., De, M.L., De Schamphelaere, K.A., 2013. The initial tolerance to sub-lethal Cd exposure is the same among ten naive pond populations of *Daphnia magna*, but their micro-evolutionary potential to develop resistance is very different. Aquat. Toxicol. 144-145, 322–331.

Messiaen, M., Janssen, C.R., Thas, O., De Schamphelaere, K.A., 2012. The potential for adaptation in a natural *Daphnia magna* population: Broad- and narrow-sense heritability of net reproductive rate under Cd stress at two temperatures. Ecotoxicology 21, 1899–1910.

Morley, N.J., 2009. Environmental risk and toxicology of human and veterinary waste pharmaceutical exposure to wild aquatic host–parasite relationships. Environ. Toxicol. Pharmacol. 27, 161–175.

Muyssen, B.T., Janssen, C.R., Bossuyt, B.T., 2002. Tolerance and acclimation to zinc of field-collected *Daphnia magna* populations. Aquat. Toxicol. 56, 69–79.

Pope, K.L., Lochmann, S.E., Young, M.K., 2010. Methods for assessing fish populations. In: Quist, M.C., Hubert, W.A. (Eds.), Inland Fisheries Management in North America, third ed. American Fisheries Society, Bethesda, MD, pp. 326–351.

Sibly, R., Calow, P., 1987. Ecological compensation: A complication for testing life-history theory. J. Theor. Biol. 125, 177–186.

Sibly, R., Calow, P., Nichols, N., 1985. Are patterns of growth adaptive? J. Theor. Biol. 112, 553–574.

Sibly, R.M., Hone, J., 2002. Population growth rate and its determinants: An overview. Philos. Trans. R. Soc. Lond. B. Biol. Sci. 357, 1153–1170.

Walker, C.H., Sibly, R.M., Hopkin, S.P., Peakall, D.B., 2012. Principles of Ecotoxicology, fourth ed. CRC Press, Boca Raton, FL.

Effects of Chemicals on Aquatic Communities and Ecosystems

Abstract

Community and ecosystem ecotoxicology are divided so that the former describes the changes in the species assemblies and their properties and the latter the changes in processes involving abiotic components, thereby including energy flow and biogeochemical cycles. Effects on communities are usually described using diversity indices, since sublethal, functional parameters, informative of most members of the community, are hard to come by. Contaminants usually decrease the species diversity of the community. The diversity can decrease either in all trophic levels or in only one or some of them. Exposure of communities to contaminants can result in increased tolerance to the contaminants. This can be evaluated as pollution-induced community tolerance (PICT). An important component of acquiring community tolerance is that sensitive species are replaced by more tolerant ones. Tolerance induced to one contaminant can result in increased tolerance to a novel contaminant, if the contaminants share similar modes of action or detoxification. On the other hand, if acclimation to a toxicant involves an increased energy cost, exposure to a novel toxicant can be associated with decreased tolerance. The speed of recovery from toxicant exposure is more rapid in a previously fluctuating than in a stable environment. Contaminants can affect virtually all processes involved in ecosystem function, thus playing an important role in present environmental problems such as eutrophication, algal blooms, and ocean acidification.

Keywords: species assemblages; metacommunity; Shannon diversity index; Brillouin diversity index; similarity indices; index of biological integrity (IBI); species richness; pollution-induced community tolerance (PICT).

17.1 INTRODUCTION

A community is an assemblage of populations of organisms in a given space. An ecosystem is the (biological) community together with its abiotic environment. Most of the abiotic factors are easily differentiated from the biological community. However, the abiotic environment also includes dead organic matter, such as remains of dead organisms and fallen leaves. Toxicants are a part of the abiotic environment before their entry to organisms, but all their effects in organisms are necessarily effects on components of biological communities. The term ecosystem is often also defined as an entity comprising a single, defined energy flow. Community and ecosystem toxicology cannot really be separated. However, in this chapter, the influences of contamination on species assemblages and consecutive indirect effects observed in organisms (and reflected in the different populations) are treated as community ecotoxicology, and toxicant effects that take into account effects on energy flow (both on biotic and abiotic components) and on abiotic cycles (such as nutrient and carbon cycles) are classified as ecosystem toxicology. Although community/ecosystem effects are the ultimate end points assessed before environmental decisions are made, studies addressing them are a minority in aquatic toxicology. One can envision three reasons for this. The first is that work in aquatic toxicology has largely carried on the traditions of classical toxicology, which is basically individual oriented. Second, effects on individuals form the basis of all ecosystem effects. There cannot be an ecosystem effect without the individuals of some species being affected. Because the species sensitivities and modes of action of many toxicants are poorly known, and one does not even know which toxicant effects on which species are decisive in terms of ecosystem function, further studies of individual functions are necessary to explain ecosystem effects. Third, the community/ecosystem studies are mostly long term, and their results are difficult to interpret unequivocally.

As with populations (metapopulation), a concept of metacommunity has been introduced. When a species assemblage behaves in most instances as an independent community, but it is possible that some of the species from two or more such communities interact, one may be talking of a metacommunity (see Figure 17.1).

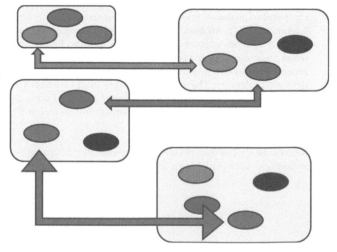

FIGURE 17.1 **The metacommunity concept.** The green rounded rectangles show different metacommunities, which are independent for most of the time. Every metacommunity consists of populations of different species (indicated as ovals; the same color of oval indicates the same species). Migration of the species between metacommunities occurs, which is indicated with arrows (having the same color as the oval of the migrating species). Since migration is possible, the different entities are not communities but metacommunities.

17.2 COMMUNITY ECOTOXICOLOGY

In community ecotoxicology, the effects of toxicants are evaluated at the community level. To do this, indices for describing biological diversity are sought. The simplest of such measures is the species richness, the number of species found in the community. However, the mere species richness does not take into account variation in the population density; some species are abundant, others very rare. A simple solution to account for this is to include only species above a certain agreed density in the species-richness data. To be more accurate, species diversity indices can be used, which weigh the population density. For ecotoxicological purposes, the most suitable ones are the Shannon and Brillouin diversity indices, since they also attach importance to rare species, which can be of importance for evaluating community effects of toxicants. The formula for the Shannon (also called the Shannon–Wiener) index is:

$$H^1 = \sum\nolimits_{t=1}^{S} p(i)\, ln\, p(i);$$ (17.1)

and the Brillouin index:

$$H = \frac{1}{N} ln \left[\frac{N!}{\prod_{(t=1)^S} n(i)!} \right]$$ (17.2)

In the first equation, S is the total number of species and $p(i)$ is the total number of individuals of species i; in the second, S is the total number of species, N is the total number of individuals, and $n(i)$ is the total number of individuals of species i. Although both estimates give similar values, the values of the Brillouin index are somewhat smaller. This is because it gives an estimate of the diversity in the sample, while the Shannon index gives an estimation of the diversity in the community. The inclusion of contributions from rare species in biodiversity indices becomes obvious from the schematic in Figure 17.2. In communities that are not impacted by toxicants, the proportion of rare species is often much higher than in toxicant-exposed communities. Now, if an index describing how the biodiversity is affected by contamination does not take the change in the contribution of rare species into account, a much reduced effect of treatment is observed. The weakness of these biodiversity indices is that they depend both on species richness and on

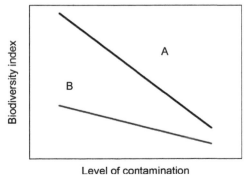

FIGURE 17.2 **Effects of contaminants on biodiversity indices, taking into account rare species and not taking them into account.** The biodiversity of the community decreases with increasing contamination. The change of value is much larger for an index taking rare species into account (A) than for an index not doing this (B).

evenness. If the treatment decreases species richness but increases evenness, no change may be seen in the index. Another very useful metric when evaluating how a community is impacted is similarity indices. They (several different indices can be found) compare the similarities between two sites (e.g. reference vs impacted site), evaluating either the presence/absence or density of the species. One can also apply the index of biological integrity (IBI), which comprises three categories: (1) species richness and the phylogenetic positions of species, (2) trophic positions of the species found, and (3) estimation of the abundance and properties (condition) of the species in the community.

In evaluating the ecotoxicological effect of contaminants on communities, it is imperative that the contribution of different trophic levels is considered. This is because many of the indirect effects of contaminants (such as predator–prey interactions, availability of shelter; see section 15.5 for a more detailed discussion of biotic interactions). Also, bioaccumulation of toxicants via food (see section 6.2) depends on trophic interactions. The species making up any community are not equally important (per capita) for the functioning of the system (Figure 17.3). Species having a much higher influence on the function of the community/ecosystem than their density would predict are called keystone species.

Generally, one is observing a decrease in the diversity of a community exposed to contamination (Figure 17.2). If, instead of determining the overall diversity indices, the population density of one species, for example, were looked at, one could see an initial increase in density, if interspecific competition and predation pressure were relieved by the reduction in the numbers/extinction of relevant species (see Figure 17.4 for a schematic example). If the community response were dissected to trophic levels, it is possible that the stress affects all trophic levels or that only one or some levels are affected (Figure 17.5). Although a decrease in the diversity (and stability) of communities is accepted as a generalization about contamination effects on them, both the determined endpoint (e.g. species-level vs community-level response) and the status of the community (equilibrium vs disequilibrium state) influence the results. The responses of a community can be markedly different for disequilibrium and equilibrium communities. Further, the equilibria that communities reach are both space and time dependent. Communities with different stabilities (equilibria) can be developed from initially similar "seeds." Consequently, although micro- and mesocosm experiments are the best alternatives for controlled community investigations, one cannot be certain, even when different replicates are stable, that they would present similar equilibrium states (for species diversities and densities; see section 5.2). One thing that needs to be noted is that, when studying the overall community responses to contamination, one can seldom use specific functional sublethal end points, since these, which could be measured and give valuable information on all members of a community, are virtually impossible to find.

Communities can be exposed to contaminants as pulses or more continuously. This affects the responses, with schematic examples given in Figure 17.6. Communities can acquire tolerance to pollutants (pollution-induced community tolerance, or PICT). The use of PICT in the assessment of community responses to contaminants has three assumptions:

1. The sensitivities of species in the community to contaminants vary.
2. Contaminants restructure communities; sensitive species are replaced by tolerant ones.
3. Differences in community tolerance can be evaluated with short-term experiments.

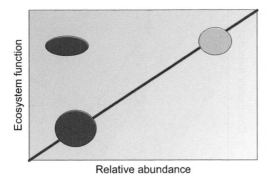

FIGURE 17.3 **The importance of species for ecosystem function.** In most cases, the importance increases with increasing proportion of the species of the community biomass, as indicated by the line on the graph. Rare species are normally of little importance (purple oval) and abundant ones highly important (pink oval). In the area above the line, a species is ecologically more important than its abundance would suggest, and in the area below the line, a species is less important for ecosystem function than its abundance would suggest. Keystone species (red oval) are species that are much more important for ecosystem function than their abundance would suggest.

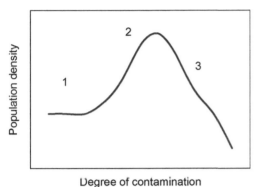

FIGURE 17.4 **Effects of contamination on the density of a species.** (1) In the absence and low concentration of contaminant, the population density is stable and partially regulated by predators and competition. (2) With increasing contaminant concentration, the densities of the more sensitive predator and competing species decrease. The decreasing predation pressure and competition allow the population density to increase. (3) A further increase in contaminant concentration causes toxicity, whereby the population density of the species decreases.

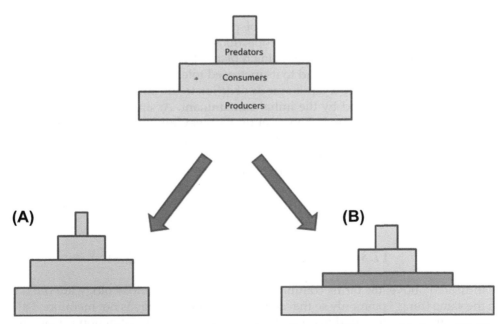

FIGURE 17.5 **A schematic representation of contaminant effects on different trophic levels.** (A) All trophic levels are affected (indicated as green color in the trophic pyramid). (B) Only one trophic level is affected (indicated by the blue rectangle in the trophic pyramid).

FIGURE 17.6 **Pulsed vs continuous exposure to a contaminant.** The blue arrow indicates a sudden pulse exposure, and the dark line the community response. Usually the community recovers from a pulse exposure. The light blue area indicates a continuous exposure to a contaminant, with the orange line indicating the community response, which is seen throughout the exposure.

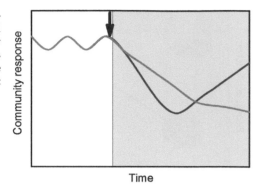

FIGURE 17.7 **Recovery from contamination in fluctuating and stable habitats.** The orange area indicates contaminant exposure. The recovery is much faster in the fluctuating environment (dark gray line) than in the stable environment (blue line).

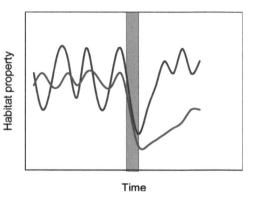

In addition to changes in the tolerance of populations in the communities, which are largely dependent on the increase in the proportions of pre-existing tolerant genotypes, PICT can be due to the more sensitive species being replaced by more tolerant ones. A response of a community to a contaminant can lead to its increased tolerance to a novel contaminant, if the new contaminant either shares a similar mode of action to the initial contaminant or is detoxified by a mechanism induced by the initial contaminant. As an example, if the initial contaminant induces the aryl hydrocarbon receptor pathway, all further contaminants using the same pathway will be detoxified more effectively as long as the induction persists. Although increased tolerance to a novel contaminant is common, communities can also respond to novel toxicants by decreased tolerance. This happens if the acclimation of the community to the contaminant carries a (energy) cost, which decreases the energy available for acclimation to a new toxicant. A further point to note is that any recovery from a contaminant is more rapid in variable than in stable habitats (Figure 17.7).

17.3 ECOSYSTEM ECOTOXICOLOGY

As indicated above, the terms community and ecosystem ecotoxicology are often used to mean the same thing, commonly so that community ecotoxicology is not mentioned at all, but community responses included in ecosystem responses. Here, a distinction is made to define

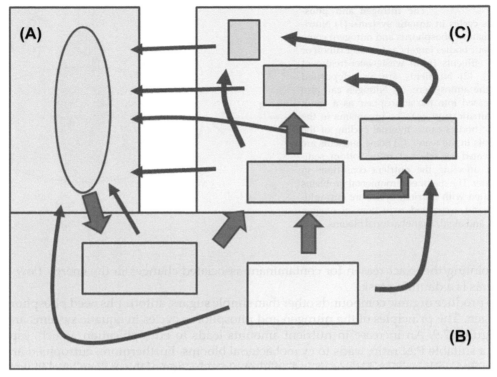

FIGURE 17.8 **A schematic representation of energy flow in an ecosystem.** (A) The oval indicates dead organic matter. (B) Producers; left-hand rectangle indicates detritivores, which transfer energy from dead organic matter to the biological community; right-hand rectangle indicates autotrophs, which translate solar energy to biomolecules that can be used to drive energy consumption by other organisms. (C) Consumers; blue rectangles indicate primary and secondary consumers; gray rectangles primary and top predators. Arrows indicate the pathways of energy transfer. Contaminants can affect virtually every step of energy flow.

a community response as a response where species alterations (including the properties of populations within species) occur, and an ecosystem response as one where energy flow or biogeochemical cycles are affected. Ecosystem responses can vary in size from small-scale ecosystem responses (e.g. what happens in a small pond) to global variations (e.g. ocean acidification).

The first type of ecosystem response involves how contamination affects energy allocation in the ecosystem. Figure 17.8 gives a very simplified picture of energy allocation in an ecosystem. Basically, energy becomes available to the biological community of the ecosystem from two sources: first, autotrophs produce organic molecules from carbon dioxide using light energy (photosynthesis), and, second, detritus-eating organisms (detritivores) return dead organic matter to the energy cycle. All the other organisms consume energy, and upon the death of both the producers and the consumers, their energy is transferred to the detritus pool, from which it is returned to the energy cycle by detritivores. Contaminants can affect the energy allocation of the ecosystem at any point. For example, any change in the efficiency of photosynthesis is seen in a change in usable energy by consumers. This being the case,

FIGURE 17.9 **The nitrogen and phosphorus cycles in aquatic systems.** (1) Nutrients (mainly phosphorus and nitrogen) come to aquatic bodies largely as fertilizer runoff or urban effluents (from wastewater-treatment plants). (2) Nutrients are also deposited from the atmosphere. (3) Nitrogen can also be released into the atmosphere as a result of denitrification. (4) The organisms in the aquatic bodies cause internal cycling of the nutrients in the water. (5) Some nutrients are sedimented or released from bottom sediments, affecting the nutrient conditions in the water. The major environmental problems associated with nutrient cycles are eutrophication and associated reductions of oxygen levels, and algal/cyanobacterial blooms.

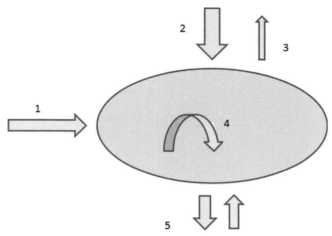

pinpointing the exact reason for contaminant-associated changes in the energy flow of ecosystems is a daunting task.

To produce organic compounds other than simple sugars, autotrophs need phosphorus and nitrogen. The principles of the nitrogen and phosphorus cycles in aquatic systems are given in Figure 17.9. An increase in nutrient amounts leads to eutrophication, which, especially with a suitable P:N ratio, leads to cyanobacterial blooms. Furthermore, eutrophied areas are associated with large variations in and an ultimate reduction of the oxygen level in water (see section 15.2). These are the major issues associated with nutrients in aquatic environments.

With regard to carbon cycling, about 85% of the earth's carbon pool exists in oceans. Further, about half of the total photosynthesis on earth is carried out by oceanic, mainly microscopic, algae. The major contamination-associated problem of the carbon cycle is ocean acidification, which can be influenced by the effects of toxicants on photosynthetic efficiency of oceanic microalgae. One such group of contaminants is compounds in oil. As a result of reduced photosynthetic efficiency of microalgae, carbon dioxide conversion to molecular oxygen is reduced, and consequently the carbon dioxide–bicarbonate–carbonate equilibrium in water is shifted to the left, and oceanic pH is reduced. The expected changes in oceanic pH will also influence the speciation of especially copper and uranium, which will affect the toxicity of these metals to oceanic organisms.

Relevant Literature and Cited References

Blanck, H., 2002. A critical review of procedures and approaches used for assessing pollution-induced community tolerance (PICT) in biotic communities. Human. Ecol. Risk Assess. 8, 1003–1034.

Borja, A., Chust, G., del Campo, A., Gonzalez, M., Hernandez, C., 2013. Setting the maximum ecological potential of benthic communities, to assess ecological status, in heavily morphologically-modified estuarine water bodies. Mar. Pollut. Bull 71, 199–208.

Burton Jr, G.A., Landrum, P.F., 2003. Toxicity of sediments. In: Middleton, G.V., Church, M.J., Corigilo, M., Hardie, L.A., Longstaffe, F.J. (Eds.), Encyclopedia of Sediments and Sedimentary Rocks. Kluwer, Dordrecht, Germany.

Camargo, J.A., Alonso, A., 2006. Ecological and toxicological effects of inorganic nitrogen pollution in aquatic ecosystems: A global assessment. Environ. Int. 32, 831–849.

Fischer, B.B., Pomati, F., Eggen, R.I., 2013. The toxicity of chemical pollutants in dynamic natural systems: The challenge of integrating environmental factors and biological complexity. Sci. Total Environ. 449, 253–259.

Hebert, C.E., Arts, M.T., Weseloh, D.V., 2006. Ecological tracers can quantify food web structure and change. Environ. Sci. Technol. 40, 5618–5623.

Jones, J.G., 2001. Freshwater ecosystems: Structure and response. Ecotoxicol. Environ. Saf. 50, 107–113.

Kuzyk, Z.Z., Macdonald, R.W., Johannessen, S.C., Stern, G.A., 2010. Biogeochemical controls on PCB deposition in Hudson Bay. Environ. Sci. Technol. 44, 3280–3285.

Legorburu, I., Rodriguez, J.G., Borja, A., Menchaca, I., Solaun, O., Valencia, V., Galparsoro, I., Larreta, J., 2013. Source characterization and spatio-temporal evolution of the metal pollution in the sediments of the Basque estuaries (Bay of Biscay). Mar. Pollut. Bull 66, 25–38.

Millero, F.J., Woosley, R., Ditrolio, B., Waters, J., 2009. Effect of ocean acidification on the speciation of metals in seawater. Oceanography 22, 72–85.

Nayar, S., Collings, G., Pfennig, P., Royal, M., 2012. Managing nitrogen inputs into seagrass meadows near a coastal city: Flow-on from research to environmental improvement plans. Mar. Pollut. Bull 64, 932–940.

Newman, M.C., Clements, W.H., 2008. Ecotoxicology: A Comprehensive Treatment. CRC Press, Boca Raton, FL.

Pascual, M., Borja, A., Franco, J., Burdon, D., Atkins, J.P., Elliott, M., 2012. What are the costs and benefits of biodiversity recovery in a highly polluted estuary? Water Res. 46, 205–217.

Pelletier, E., Sargian, P., Payet, J., Demers, S., 2006. Ecotoxicological effects of combined UVB and organic contaminants in coastal waters: A review. Photochem. Photobiol. 82, 981–993.

Sandford, R.C., Exenberger, A., Worsfold, P.J., 2007. Nitrogen cycling in natural waters using in situ, reagentless UV spectrophotometry with simultaneous determination of nitrate and nitrite. Environ. Sci. Technol. 41, 8420–8425.

Sibley, P.K., Harris, M.L., Bestari, K.T., Steele, T.A., Robinson, R.D., Gensemer, R.W., Day, K.E., Solomon, K.R., 2001. Response of phytoplankton communities to liquid creosote in freshwater microcosms. Environ. Toxicol. Chem. 20, 2785–2793.

Sparrevik, M., Saloranta, T., Cornelissen, G., Eek, E., Fet, A.M., Breedveld, G.D., Linkov, I., 2011. Use of life cycle assessments to evaluate the environmental footprint of contaminated sediment remediation. Environ. Sci. Technol. 45, 4235–4241.

Wang, J., Pei, Y.S., Zhang, K.J., Cao, G., Yang, Z.F., 2011. Investigating the spatial-temporal variation of nitrogen cycling in an urban river in the North China Plain. Water Sci. Technol. 63, 2553–2559.

Zepp, R.G., Erickson III, D.J., Paul, N.D., Sulzberger, B., 2011. Effects of solar UV radiation and climate change on biogeochemical cycling: Interactions and feedbacks. Photochem. Photobiol. Sci. 10, 261–279.

Abstract

Two reasons make model building an important part of aquatic toxicology. First, for making environmental decisions, one must be able to predict how contaminants will affect aquatic ecosystems in the future. This can only be done if the results of retrospective or laboratory studies can be modeled to give plausible future scenarios. However, any models are only as appropriate as their weakest components. Second, the multitude of both species and compounds in ecosystems requires that models are developed that take both into account. The major interplay between ecotoxicological studies and environmental management is in risk assessment, which uses the results of scientific studies to evaluate risks that contaminants cause to the environment. The widest range of data regarding both different species and different toxicants is available on lethality, especially in the form of LC50 (lethal concentration for 50% of organisms) values. Lethality data are often analyzed using probit analysis. Chemical structures are related to their functions using the quantitative structure–activity relationship (QSAR). Its effective use requires that the system and end points used are toxicologically well characterized, and specific to the chemical structures evaluated. Most of the modeling of uptake, metabolism, and excretion (toxicokinetic modeling) uses operational compartments, because the treatment of physiologically based models becomes quite complex. An aspect of uptake relates to the bioavailability of the compounds, which can be modeled, especially for metals in freshwater. Regulatory agencies use, in particular, the biotic ligand model (BLM) for modeling metal uptake and effects.

Keywords: risk assessment; mode of action; hazard; exposure; fugacity; predicted no effect concentration (PNEC); predicted environmental concentration (PEC); risk quotient; probit analysis; acute-to-chronic toxicity ratio (ACR); Kaplan–Meier estimator; lethal concentration 50 (LC50); lethal dose 50 (LD50); no observed effect concentration (NOEC); lowest effect concentration (LOEC); internal effective concentration (IEC); physiologically based toxicokinetic (PBTK) model; biotic ligand model (BLM); Windermere humic aqueous model (WHAM).

18.1 INTRODUCTION

The goal of aquatic toxicology is to predict the effects of contaminants in ecosystems. This should form the basis of policies affecting aquatic systems (see Figure 18.1 for a scheme of how environmental management an toxicological studies should interact). Predictions require that existing observations can be used to generate scenarios about the future. This is the case, although aquatic toxicological studies in the natural environment necessarily report what has already happened (thus being retrospective). This holds also for biomonitoring studies, which rely on it being possible to extrapolate from any observed trend to the future (see Figure 18.2). The utility of retrospective studies is twofold. First, they can indicate how

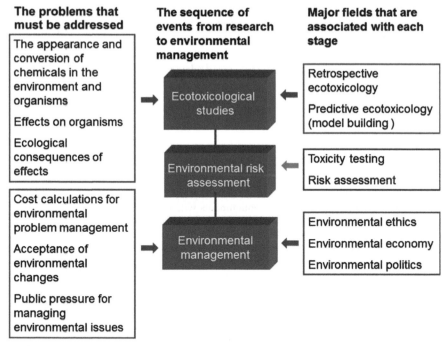

FIGURE 18.1 **A scheme of how environmental management follows from ecotoxicological studies.** The same color in arrows and box outlines indicates the same stage in the decision-making pathway. Note that toxicity testing and risk assessment are not included in ecotoxicological research. For toxicity testing, this is because testing uses standard methods with the aim of making comparisons. Development of new testing methods is part of ecotoxicological research. Risk assessment itself is not research, but uses the data gathered in ecotoxicological studies to evaluate the risks that the contaminant(s) assessed cause to the environment.

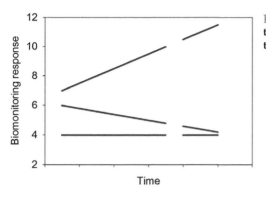

FIGURE 18.2 **Extrapolation from biomonitoring data to the future supposes that the trends measured (lines before the break) continue in the future (lines after the break).**

the environment has already been affected. Such information can be required in order to identify the sources of contamination, to require those responsible either to carry out stricter cleaning of their effluent or to pay for the damages incurred. Both necessitate the damaging contamination being pinpointed to a definite source, which requires highly specific exposure biomarkers. Second, they can be used to prevent similar discharges and effects in other places. This requires the assumption that other environmental factors do not significantly affect the contaminant responses that are observed. Laboratory studies, on the other hand, can use novel contaminants and their mixtures. Thus, possible contaminants can be studied before their appearance in the environment. However, these studies do not include all the contaminants, their interactions, and interactions with natural abiotic and biotic factors. Consequently, one must always assume that the factors studied in the laboratory are decisive for the effects in nature.

From the above it is clear that building and using models is necessary for predictive aquatic toxicology. A flowchart of model building is given in Figure 18.3. It should be noted that any model is as good (or bad) as its least accurate component. A factor complicating any predictions is that the behavior of chemicals in the aquatic environment is highly complex. Figure 18.4 summarizes the aspects that need to be taken into account when building individual-based models. The toxicants may, further, have different targets (modes of action, MOA) at different concentrations (see Figure 18.5). This heterogeneity has been poorly included in modeling of toxicant effects done so far.

18.2 RISK ASSESSMENT

In risk assessment, one seeks to evaluate how likely an adverse effect (hazard) is and what its consequences are. This generally consists of (1) hazard identification, (2) hazard characterization, (3) exposure assessment, and (4) risk characterization.

18.2.1 Hazard Identification

The first step of risk assessment is the identification of hazard. In aquatic toxicology, this means that a chemical known to cause effects based either on laboratory experiments or on

FIGURE 18.3 **Flowchart of model building.** In model building, one has to evaluate whether data are available for all the components needed for the model. If not, a research question and consequent experiment to address the lack of information is required.

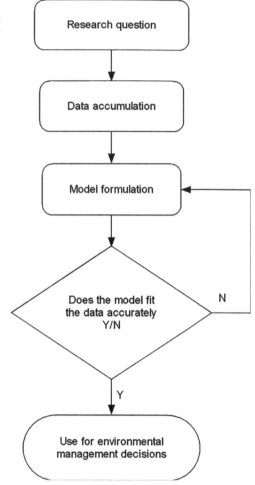

field information from other places is released into the environment. Thus, chemical measurements are important at this point. Also, in identifying a hazard in the natural environment, one should be able to identify the most probable species/groups affected by the contaminant. This is not an easy task given the diversity of organisms and their large sensitivity differences to many chemicals. The most detailed information available on many species is about acute lethal toxicity. Consequently, one must assume that the most likely species/groups are the ones for which the chemical shows the greatest acute lethal toxicity. Identification of hazard should also take into account the possibility and probability of chemical interactions in water, i.e. the likelihood of both agonistic and antagonistic interactions between chemicals occurring. Further, one needs to be able to determine whether the effects of toxicants are additive, since in natural environments chemicals occur in complex cocktails. One also needs to be able to determine the bioavailable fraction of the chemical. Thus, although in principle the identification of hazard is a straightforward evaluation if potentially toxic chemicals reach the environment, the many qualifying aspects make this identification a complex task.

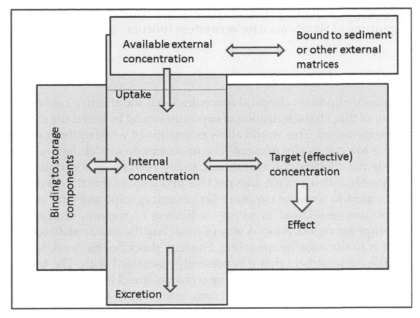

FIGURE 18.4   In building models about toxicant effects on organisms, one needs to take into account bioavailability (green rectangle), build toxicokinetic models (pink rectangle), and consider the organismic distribution and its influence on toxicant effects (blue rectangle).

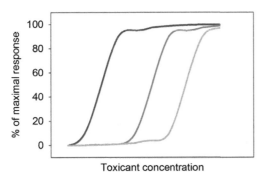

FIGURE 18.5   A toxicant may have several modes of action (targets), varying with concentration; different concentration–response profiles are depicted with lines of different colors.

18.2.2 Hazard Characterization

Characterization of a hazard is probably the most straightforward part of risk assessment, since it involves the determination of the most likely target of toxicity in an organism. Since most aquatic organisms have not been studied in any detail, one can use the "read-across hypothesis," which supposes that the mode of action of a chemical is the same in different organisms as long as they have appropriate structures for the action. While this is quite a sensible point of departure, one needs to remember that it is also possible that homologous proteins or pathways assume different functions. Another point to remember is that even when the function of a protein remains the same, its affinity for any ligands may vary markedly.

When doing risk assessment in aquatic ecosystems, in hazard characterization one must evaluate which organisms of significance for ecosystem function are most likely to be at risk.

18.2.3 Assessment of Exposure

The toxicologically important chemical concentration is the effective concentration in the target tissue. In view of this, characterization of exposure would be best if the concentration in the target tissue were measured. This would allow estimation of whether the concentration caused by the exposure is toxicologically relevant. The toxicokinetic models (see section 18.4) seek to determine the internal effective concentration. Usually, nominal concentration is not adequate for exposure characterization, since it does not take into account that the chemical can be bound to compartments inert in terms of exposure (in laboratory exposures, for example, tubing or walls of the exposure vessel, and in nature, sediments or surfaces of organisms). Also, the chemicals can escape by vaporization. A way of modeling the steady-state concentration in the phase of interest is to use fugacity modeling. Fugacity describes the tendency of a compound to escape from the compartment that it is presently associated with. The major advantage of using fugacity-based modeling when exposure is characterized is that a single metric (and units mol/volume) can be used for all the different compartments. To facilitate comparisons between toxicants, whenever possible their concentrations should be given as molar concentrations, as this enables instantaneous comparisons between chemicals with different molecular weights. Whenever possible, one should measure both the actual environmental concentration and the effective concentration in the target compartment to characterize exposure.

18.2.4 Characterization of Risk

When one knows what the probable hazard is (and the concentration dependence of adverse effect; predicted no effect (environmental) concentration, PNEC), and the predicted environmental concentration (PEC), one can calculate the so-called risk quotient (RQ):

$$RQ = PEC/PNEC. \tag{18.1}$$

The PNEC is usually calculated from LC50 values (lethal concentration 50, the concentration lethal to 50% of organisms), which are divided by arbitrarily defined constants (10–100–1000). The magnitude of the constant depends on the level of knowledge about the system: the better the system is known, the smaller the magnitude of the constant. There are two reasons for using the uncertainty constants. The first is that the acute LC50 is much higher than the concentrations causing effects in ecosystems. Second, it is not likely that the most sensitive species of the ecosystem would have an LC50 value available. In risk assessment, one follows the precautionary principle.

18.3 MODELS WITH LETHALITY AS AN END POINT

Traditionally, the end point of toxicological studies has been death. This end point can be unequivocally determined only with vertebrates (and in invertebrates in which the cessation of heartbeat can be determined), so in invertebrates and plants, end points such as immobility and cessation of growth or of cell division are taken as equivalent end points. In most cases, the methodology has been directly transferred from mammalian toxicology to aquatic toxicology. Thus, it is customary to talk about "dose" when, in fact, dissolved toxicant concentration in the ambient water is known. It is not known what dose (amount of toxicant reaching the animal) one actually has, if the amount in the animal has not been measured. For the same concentration in the ambient water, the amount going into an animal can be markedly different, depending on many factors determining the uptake of toxicant in the animal. Consequently, one should talk about concentration–response relationships when contaminants dissolved in aquatic systems are talked about. Doses should be talked about when either the contaminant enters the animal in food or is injected (or if one is talking about radiation dose).

When lethality is considered as the end point, the most common treatment of data involves probit transformation. The probit transformation is derived from the normal equivalent deviation, the proportion of organisms dying expressed as standard deviations from the mean of the normal curve. Probit transformation yields a straight line for the concentration (dose)–response relationship. Obtaining a straight line was necessary before the advent of effective personal computers, when any analysis needed to be done on graph paper. Notably, probit analysis graph paper was commonly utilized. Instead of doing the probit transformation, it is possible to carry out logit transformation. The result is very similar to that obtained from probit transformation. Tools for doing probit analysis are included in most statistical packages, and can also be freely downloaded from the internet.

When calculating lethality, the most commonly used end point is the median lethal concentration (dose) (i.e. 50% of the organisms die; LC50, LD50). This is used especially because in the middle of the distribution, the variance is smallest, whereby the variation of concentrations causing a certain degree of lethality at 10–90% range is small. In addition, commonly used values are lowest effect concentration (LOEC) and the highest no observed effect concentration (NOEC). Both are similar values and normally estimated from the probit lines. The lethality is strongly dependent on time, as illustrated in Figure 18.6. The lethality approaches asymptotically some limiting value that can be described as the chronically lethal concentration. One parameter relating to lethal toxicity is the acute-to-chronic toxicity ratio (ACR). Because of the time dependency of lethal toxicity, ACR depends on the length of the exposure period that is taken to represent acute toxicity. Another complicating factor affecting lethality evaluations is that, depending on the mechanism of toxicity, it is entirely possible that no mortality occurs during the exposure (e.g. 24 h), but significant mortality

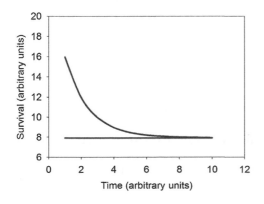

FIGURE 18.6 **Time dependency of survival of toxicant exposure.** As indicated by the curved blue line, survival decreases with time, asymptotically approaching the limiting survival value (chronic lethality), indicated as a red line.

FIGURE 18.7 **The mortality caused by three different con-taminants, analyzed using probit analysis.** The probit lines (black, red, and green) yield the same LC50 value, but different NOEC and LOEC values. The different slopes of the probit lines indicate different concentration dependence of mortality for the different toxicants.

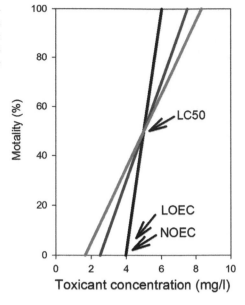

FIGURE 18.8 **An example of a Kaplan–Meier plot.** The plot relates survival and exposure time. The figure gives time-to-death (survival) data for two toxicants (black and red lines).

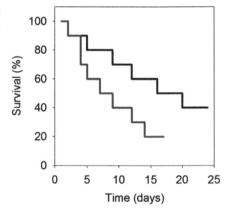

is observed after the organisms are returned to clean water: should such mortality, observed, for example, after a week, be included in the 24-h mortality, as it is caused by the 24-h toxicant exposure? The differences of chemicals in terms of lethality are seen from the slope of probit lines (Figure 18.7). The steeper the line, the smaller the difference between the NOEC and the concentration causing 100% mortality. Another way of looking at mortality caused by a toxicant is to estimate the time to death. A commonly used method is the Kaplan–Meier (product limit) estimator. Figure 18.8 gives an example of a Kaplan–Meier plot. Commercially available statistical packages enable the use of these metrics in toxicity evaluations.

A problem relevant to the use of lethality in estimating how toxicants affect ecosystem function is to be able to relate mortality in the lethality tests with limited time to ecosystem effects in nature, where toxicants have usually exerted their effect for a much longer, but often unknown, time. Usually, the concentrations of toxicants found in nature are much lower than the ones causing mortality in short-term toxicity tests. Another important point is to ascertain that one has data on species that influence the ecosystem function or, if such data are not available, data on another species that behaves similarly to the ecologically relevant species.

18.4 TOXICOKINETIC MODELING

Toxicokinetic models integrate the uptake, metabolism, and excretion of chemicals in an individual organism (see Figure 18.4). The major aim of the modeling is to get the internal effective concentration (IEC), which is usually synonymous to critical body residue, CBR). In the simplest case, the organism is considered as a single compartment, and both the uptake and the excretion of chemicals can be thought of as simple exponential uptake and loss. In reality, some of the compartments in an organism are inert and others are targets for the toxic action of chemicals. Mathematical modeling of multicompartment toxicokinetics is more complex than that of single compartment behavior, but still very much simplified from the real situation. The compartmental models require only limited time-series data for a compound in a reference compartment to allow estimation of pharmacokinetic parameters, which characterize the uptake of the chemical from water, its distribution, and its half-life (half-life takes into account the breakdown of the chemical and its excretion). The compartments in the models are strictly operational, and do not represent any physiological phenomena. In physiologically based toxicokinetic (PBTK) models, the compartments are chosen so that they represent physiologically relevant entities (Figure 18.9). The data are collected from different tissues/physiological compartments, and their perfusion is taken into account. While such models increase the realism of toxicokinetics in aquatic organisms, they are increasingly complex, whereby their mathematical treatment is not within the reach of most aquatic toxicologists.

18.5 QSAR

The quantitative structure–activity relationship (QSAR) is based on the assumption that the activity of a molecule is related to its structure, so that similar molecules have similar activities. Because of the thousands of chemicals that enter the environment, QSARs are frequently used in aquatic (and other environmental) toxicology, especially in chemical regulation and risk assessment. Although the assumption of similar structure–similar activity holds true in most cases, there are exceptions. These occur especially in cases where the molecule is a specific ligand of a protein and cannot be substituted in the active site by structurally similar molecules. The QSAR approach is used in a regulatory context, e.g. when priorities are attached for chemical testing, and when risks of different chemicals are assessed. As for modeling in general, the QSAR is only as good as the data and biological knowledge it is based upon. For example, to be relevant for a QSAR, the toxicological effect must be caused by the compound itself and not by its metabolites. A commonly used end point in QSAR evaluations is lethality. Since death is caused by everything that is severe enough, it is not very specific for any structure.

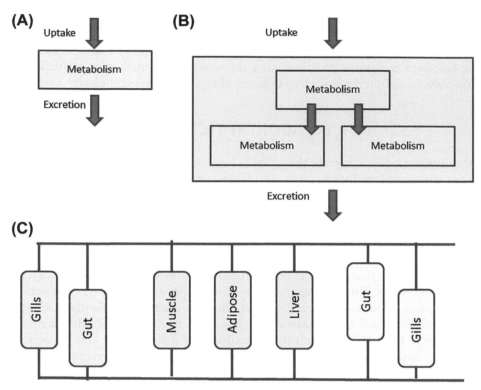

FIGURE 18.9 **Toxicokinetic models.** (A) The single-compartment model. (B) The multicompartment model. In multicompartment models, the calculations of uptake and excretion become complex. The different compartments are defined operationally and have no preconceived notions about physiological relationships. (C) Physiologically based toxicokinetic (PBTK) models divide uptake, metabolism, and excretion (signified in the figure by different colors) into functionally relevant compartments. The perfusion of different compartments is also taken into account. These factors make the models, and their mathematical treatments, quite complex.

18.6 MODELING THE PROPERTIES OF WATER AFFECTING TOXICITY (OF METALS)

The relationship between the total concentration of chemicals and the bioavailable fraction has been best modeled for metals. Variations in pH; in concentrations of cations such as calcium, magnesium, and sodium; in alkalinity; and in the presence of natural organic matter can all have a significant effect on metal toxicity. Consequently, the toxicity of a given metal can vary markedly in different water bodies at a constant total metal concentration. This presents a difficult problem for establishing regulatory guidelines, as unless the effects of different factors on metal toxicity can be predicted, a conservative limit (high concentration) is typically selected for a regulatory limit of metals in water. Two major approaches are available for evaluating how water quality affects metal toxicity: the biotic ligand model (BLM) and the WHAM-FTOX model. The latter describes the combined toxic effects of protons and metal cations towards aquatic organisms through the toxicity function (FTOX), which takes into account the binding of the metal to humic acid, calculated using the WHAM chemical speciation model. WHAM (the Windermere humic aqueous model) also takes into account the precipitation of aluminum and iron oxides, cation exchange of clay minerals in bottom sediments, and adsorption–desorption reactions of ions to fulvic acid. The model emphasizes the importance of ion accumulation to the diffuse layers surrounding the humic molecules. The WHAM program can be bought from http://www.ceh.ac.uk/products/software/wham/index.html. BLM tries to estimate how the metal ion accumulates in the gills of organisms and is transported through the gill epithelium. Because tissues for metal uptake are somewhat different in marine and freshwater environments (see section 15.3), accurate estimations using BLM are currently restricted to the freshwater environment.

Relevant Literature and Cited References

Ashauer, R., Boxall, A., Brown, C., 2006. Predicting effects on aquatic organisms from fluctuating or pulsed exposure to pesticides. Environ. Toxicol. Chem. 25, 1899–1912.

Baird, D.J., Brown, S.S., Lagadic, L., Liess, M., Maltby, L., Moreira-Santos, M., Schulz, R., Scott, G.I., 2007. In situ-based effects measures: Determining the ecological relevance of measured responses. Integr. Environ. Assess. Manag. 3, 259–267.

Baird, D.J., Burton, G.A., Culp, J.M., Maltby, L., 2007. Summary and recommendations from a SETAC Pellston Workshop on in situ measures of ecological effects. Integr. Environ. Assess. Manag. 3, 275–278.

Bianchini, A., Bowles, K.C., 2002. Metal sulfides in oxygenated aquatic systems: Implications for the biotic ligand model. Comp. Biochem. Physiol. C. Toxicol. Pharmacol. 133, 51–64.

Boudreau, T.M., Wilson, C.J., Cheong, W.J., Sibley, P.K., Mabury, S.A., Muir, D.C., Solomon, K.R., 2003. Response of the zooplankton community and environmental fate of perfluorooctane sulfonic acid in aquatic microcosms. Environ. Toxicol. Chem. 22, 2739–2745.

Brack, W., Klamer, H.J., de Lopez, A.M., Barcelo, D., 2007. Effect-directed analysis of key toxicants in European river basins: A review. Environ. Sci. Pollut. Res. Int. 14, 30–38.

Bradbury, S.P., 1995. Quantitative structure–activity relationships and ecological risk assessment: An overview of predictive aquatic toxicology research. Toxicol. Lett. 79, 229–237.

Bradbury, S.P., Russom, C.L., Ankley, G.T., Schultz, T.W., Walker, J.D., 2003. Overview of data and conceptual approaches for derivation of quantitative structure-activity relationships for ecotoxicological effects of organic chemicals. Environ. Toxicol. Chem. 22, 1789–1798.

Brausch, J.M., Connors, K.A., Brooks, B.W., Rand, G.M., 2012. Human pharmaceuticals in the aquatic environment: A review of recent toxicological studies and considerations for toxicity testing. Rev. Environ. Contam. Toxicol. 218, 1–99.

Chappaz, A., Curtis, P.J., 2013. Integrating empirically dissolved organic matter quality for WHAM VI using the DOM optical properties: A case study of Cu-Al-DOM interactions. Environ. Sci. Technol. 47, 2001–2007.

Clements, R.G., Nabholz, J.V., Zeeman, M.G., Auer, C.M., 1995. The application of structure–activity relationships (SARs) in the aquatic toxicity evaluation of discrete organic chemicals. SAR QSAR Environ. Res. 3, 203–215.

Connon, R.E., Geist, J., Werner, I., 2012. Effect-based tools for monitoring and predicting the ecotoxicological effects of chemicals in the aquatic environment. Sensors (Basel) 12, 12741–12771.

Crane, M., Watts, C., Boucard, T., 2006. Chronic aquatic environmental risks from exposure to human pharmaceuticals. Sci. Total Environ. 367, 23–41.

De Laender, F., De Schamphelaere, K.A., Vanrolleghem, P.A., Janssen, C.R., 2008. Validation of an ecosystem modelling approach as a tool for ecological effect assessments. Chemosphere 71, 529–545.

De Laender, F., De Schamphelaere, K.A., Vanrolleghem, P.A., Janssen, C.R., 2009. Comparing ecotoxicological effect concentrations of chemicals established in multi-species vs. single-species toxicity test systems. Ecotoxicol. Environ. Saf. 72, 310–315.

De Laender, F., Olsen, G.H., Frost, T., Grosvik, B.E., Grung, M., Hansen, B.H., Hendriks, A.J., Hjorth, M., Janssen, C.R., Klok, C., Nordtug, T., Smit, M., Carroll, J., Camus, L., 2011. Ecotoxicological mechanisms and models in an impact analysis tool for oil spills. J. Toxicol. Environ. Health A. 74, 605–619.

Di Toro, D.M., Allen, H.E., Bergman, H.L., Meyer, J.S., Paquin, P.R., Santore, R.C., 2001. Biotic ligand model of the acute toxicity of metals: 1, Technical basis. Environ. Toxicol. Chem. 20, 2383–2396.

Dyer, S., St J Warne, M., Meyer, J.S., Leslie, H.A., Escher, B.I., 2011. Tissue residue approach for chemical mixtures. Integr. Environ. Assess. Manag. 7, 99–115.

Escher, B.I., Ashauer, R., Dyer, S., Hermens, J.L., Lee, J.H., Leslie, H.A., Mayer, P., Meador, J.P., Warne, M.S., 2011. Crucial role of mechanisms and modes of toxic action for understanding tissue residue toxicity and internal effect concentrations of organic chemicals. Integr. Environ. Assess. Manag. 7, 28–49.

Escher, B.I., Hermens, J.L., 2002. Modes of action in ecotoxicology: Their role in body burdens, species sensitivity, QSARs, and mixture effects. Environ. Sci. Technol. 36, 4201–4217.

Gross, M., Daginnus, K., Deviller, G., de Wolf, W., Dungey, S., Galli, C., Gourmelon, A., Jacobs, M., Matthiessen, P., Micheletti, C., Nestmann, E., Pavan, M., Paya-Perez, A., Ratte, H.T., Safford, B., Sokull-Kluttgen, B., Stock, F., Stolzenberg, H.C., Wheeler, J., Willuhn, M., Worth, A., Comenges, J.M., Crane, M., 2010. Thresholds of toxicological concern for endocrine active substances in the aquatic environment. Integr. Environ. Assess. Manag. 6, 2–11.

Guillen, D., Ginebreda, A., Farre, M., Darbra, R.M., Petrovic, M., Gros, M., Barcelo, D., 2012. Prioritization of chemicals in the aquatic environment based on risk assessment: Analytical, modeling and regulatory perspective. Sci. Total Environ. 440, 236–252.

Heugens, E.H., Hendriks, A.J., Dekker, T., van Straalen, N.M., Admiraal, W., 2001. A review of the effects of multiple stressors on aquatic organisms and analysis of uncertainty factors for use in risk assessment. Crit. Rev. Toxicol. 31, 247–284.

Hutchinson, T.H., Pickford, D.B., 2002. Ecological risk assessment and testing for endocrine disruption in the aquatic environment. Toxicology 181–182, 383–387.

Janssen, C.R., Heijerick, D.G., De Schamphelaere, K.A., Allen, H.E., 2003. Environmental risk assessment of metals: Tools for incorporating bioavailability. Environ. Int. 28, 793–800.

Jha, A.N., 2004. Genotoxicological studies in aquatic organisms: An overview. Mutat. Res. 552, 1–17.

Johnson, A.C., Ternes, T., Williams, R.J., Sumpter, J.P., 2008. Assessing the concentrations of polar organic microcontaminants from point sources in the aquatic environment: Measure or model? Environ. Sci. Technol. 42, 5390–5399.

Jones, D., Domotor, S., Higley, K., Kocher, D., Bilyard, G., 2003. Principles and issues in radiological ecological risk assessment. J. Environ. Radioact. 66, 19–39.

Kubinyi, H., 2008. QSAR: Hansch Analysis and Related Approaches. Wiley, Hoboken, NJ.

Lemly, A.D., 1996. Winter stress syndrome: An important consideration for hazard assessment of aquatic pollutants. Ecotoxicol. Environ. Saf. 34, 223–227.

Leonard, E.M., Marentette, J.R., Balshine, S., Wood, C.M., 2014. Critical body residues, Michaelis–Menten analysis of bioaccumulation, lethality and behaviour as endpoints of waterborne Ni toxicity in two teleosts. Ecotoxicology 23, 147–162.

Lyne, T.B., Bickham, J.W., Lamb, T., Gibbons, J.W., 1992. The application of bioassays in risk assessment of environmental pollution. Risk Anal. 12, 361–365.

McKim, J.M., Bradbury, S.P., Niemi, G.J., 1987. Fish acute toxicity syndromes and their use in the QSAR approach to hazard assessment. Environ. Health Perspect. 71, 171–186.

Moore, D.R., Breton, R.L., MacDonald, D.B., 2003. A comparison of model performance for six quantitative structure-activity relationship packages that predict acute toxicity to fish. Environ. Toxicol. Chem. 22, 1799–1809.

Munns Jr., W.R., Berry, W.J., Dewitt, T.H., 2002. Toxicity testing, risk assessment, and options for dredged material management. Mar. Pollut. Bull 44, 294–302.

Newman, M.C., 2012. Quantitative Ecotoxicology, second ed. CRC Press, Boca Raton, FL.

Niogi, S., Wood, C.M., 2004. Biotic ligand model, a flexible tool for developing site-specific water quality guidelines for metals. Environ. Sci. Technol. 38, 6177–6192.

Paquin, P.R., Gorsuch, J.W., Apte, S., Batley, G.E., Bowles, K.C., Campbell, P.G.C., Delos, C.G., Di Toro, D.M., Dwyer, R.L., Galvez, F., Gensemer, R.W., Goss, G.G., Hogstrand, C., Janssen, C.R., McGeer, J.C., Naddy, R.B., Playle, R.C., Santore, R.C., Schneider, U., Stubblefield, W.A., Wood, C.M., Wu, K.B., 2002. The biotic ligand model: A historical overview. Comp. Biochem. Physiol. C.: Toxicol. Pharmacol. 133, 3–35.

Paredes, I., Rietjens, I.M., Vieites, J.M., Cabado, A.G., 2011. Update of risk assessments of main marine biotoxins in the European Union. Toxicon 58, 336–354.

Peters, K., Bundschuh, M., Schafer, R.B., 2013. Review on the effects of toxicants on freshwater ecosystem functions. Environ. Pollut. 180, 324–329.

Rand, G.M. (Ed.), 1995. Fundamentals of Aquatic Toxicology: Effects, Environmental Fate and Risk Assessment. CRC Press, Boca Raton, FL.

Saouter, E., Pittinger, C., Feijtel, T., 2001. Aquatic environmental impact of detergents: From simple to more sophisticated models. Ecotoxicol. Environ. Saf. 50, 153–159.

Schulz, R., 2004. Field studies on exposure, effects, and risk mitigation of aquatic nonpoint-source insecticide pollution: A review. J. Environ. Qual. 33, 419–448.

Stadnicka, J., Schirmer, K., Ashauer, R., 2012. Predicting concentrations of organic chemicals in fish by using toxicokinetic models. Environ. Sci. Technol. 46, 3273–3280.

TenBrook, P.L., Tjeerdema, R.S., Hann, P., Karkoski, J., 2009. Methods for deriving pesticide aquatic life criteria. Rev. Environ. Contam. Toxicol. 199, 19–109.

Tipping, E., 2002. Cation Binding by Humic Substances. Cambridge University Press, Cambridge, UK.

Van Leeuwen, K., Schultz, T.W., Henry, T., Diderich, B., Veith, G.D., 2009. Using chemical categories to fill data gaps in hazard assessment. SAR QSAR Environ. Res. 20, 207–220.

Van Noort, P.C., Koelmans, A.A., 2012. Nonequilibrium of organic compounds in sediment-water systems: Consequences for risk assessment and remediation measures. Environ. Sci. Technol. 46, 10900–10908.

Veldhoen, N., Ikonomou, M.G., Helbing, C.C., 2012. Molecular profiling of marine fauna: Integration of omics with environmental assessment of the world's oceans. Ecotoxicol. Environ. Saf. 76, 23–38.

Veltman, K., Huijbregts, M.A., Hendriks, A.J., 2010. Integration of biotic ligand models (BLM) and bioaccumulation kinetics into a mechanistic framework for metal uptake in aquatic organisms. Environ. Sci. Technol. 44, 5022–5028.

Winter, M.J., Owen, S.F., Murray-Smith, R., Panter, G.H., Hetheridge, M.J., Kinter, L.B., 2010. Using data from drug discovery and development to aid the aquatic environmental risk assessment of human pharmaceuticals: Concepts, considerations, and challenges. Integr. Environ. Assess. Manag. 6, 38–51.

Wood, C.M., Al-Reasi, H.A., Smith, D.S., 2011. The two faces of DOC. Aquat. Toxicol. 105, suppl. 3–8.

Zeeman, M., Auer, C.M., Clements, R.G., Nabholz, J.V., Boethling, R.S., 1995. U.S. EPA regulatory perspectives on the use of QSAR for new and existing chemical evaluations. SAR QSAR Environ. Res. 3, 179–201.

Glossary

ABC transporters (ATP binding cassette transporters)   in eukaryotes active efflux transporters, which extrude toxicants from cells.

Absorption   uptake of a compound into an organism.

Acclimation   the organism familiarizes to prevailing ambient conditions without experimenter's intervention.

Acclimatization   the organism is familiarized to given conditions (involves experimenter's input).

Acetylcholinesterase (AChE)   an enzyme active in the neuromuscular synaptic cleft, which catalyzes the breakdown of acetylcholine, the most common neurotransmitter of neuromuscular junctions. Many insecticides affect acetylcholinesterase function.

Active transport   transport of compounds across membranes. Active transport uses energy directly (ATP is normally converted to ADP to make the transport possible). A well-characterized example is the sodium pump (Na^+, K^+ ATPase), which generates and maintains uneven distributions of sodium and potassium between the extra- and intracellular compartments.

Acute lethality   the toxicant-induced death occurs acutely. Commonly, the lethality is considered acute if it occurs in less than 96 h. Notably, the term acute should take the life length/generation time of the organism into account.

Adenylate energy charge   ([ATP] + 1/2[ADP]) / ([ATP] + [ADP] + [AMP]).

Adsorption   accumulation of a compound onto a particle or organism. In adsorption the accumulated compound remains on the surface and is not taken in.

Aerobic digestion   treatment of sludge with aerobic bacteria to reduce oxygen demand (see Chapter 3).

Aerobic scope   the activity of an organism that can occur aerobically.

Agonism (potentiation)   the toxicity of a compound is increased by a second compound (see Chapter 13).

AhR (aryl hydrocarbon receptor, dioxin receptor)   a bHLH-PAS-group transcription factor (bHLH = basic helix–loop–helix; PAS = PAS domain, which is the first letters of the three proteins first found with this domain, i.e. period, ARNT, and simple minded) to which many polyaromatic hydrocarbons bind, whereafter a xenobiotically activated transcription pathway is induced.

Aliphatic hydrocarbon   hydrocarbon structure without the presence of rings.

Alkalinity   the relative degree of basic property of a solution on the acid–base scale.

Allele   one of a pair of genes that occupy the same position on a chromosome. The two alleles on the chromosomes of diploid organisms separate in cell division.

Allometry   the effect of size on structures and functions.

Alpha radiation   radioactive emission consisting of helium nuclei.

Anaerobic digestion   treatment of sludge with anaerobic bacteria to reduce oxygen demand (see Chapter 3).

Androgen receptor   nuclear receptor binding male reproductive hormones and their mimics.

Anoxia   complete lack of oxygen.

Anthropogenic   caused by human activities.

Apoptosis   programmed cell death. Orderly cell death, induced by many contaminants. The need for apoptosis can be signaled by death receptors. The death signaling leading to apoptosis usually includes caspase enzymes. In apoptosis the cell first shrinks, and is then broken up to apoptotic bodies which can be phagocytosed. In addition to being induced by contaminants, apoptosis is needed for normal embryonic development and elimination of damaged or superfluous cells (see section 11.8).

Aromatic hydrocarbon (aryl hydrocarbon)   hydrocarbons containing benzene ring(s).

ARNT (aryl hydrocarbon receptor nuclear translocator)   the dimerization partner of aryl hydrocarbon receptor (AhR) and hypoxia-inducible factor alpha (HIF alpha). The dimer binds to the XRE (xenobiotic response element) or HRE (hypoxia response element) in the promoter region of the target gene and induces transcription.

Assimilation efficiency   describes how effectively nutrients and micronutrients are utilized in anabolism.

Behavioral toxicology   toxicant effects are observed in the behavior of organisms.

Beta radiation   radioactive emission in which the radiation consists of electrons or positrons.

Bioaccumulation   describes how a chemical is taken up by organisms (see Chapter 6).

Bioaccumulation factor (BAF, BF)   gives a numerical value to the accumulation of a chemical in an organism. $BF = C_{organism}/C_{source}$.

Bioamplification   indicates if the organism takes up a compound above the level occurring in the environment.

Bioavailability   indicates how well a compound can be taken up by organisms. For example, if a compound is tightly associated with large particles, its bioavailability is small.

Bioconcentration   describes how a compound is concentrated in organisms.

Bioconcentration factor (BCF)   gives a numerical value to the accumulation of a compound in organisms.

Biofilm   community of microorganisms and protists (including prokaryotes, unicellular algae, fungi, and protozoans) developing on solid surfaces in aquatic systems.

Biological half-life   numerical estimation of 50% of the time that a compound persists biologically.

Biomagnification   a term that describes how the concentration of a compound increases along the trophic chain.

Biomagnification factor   gives a numerical value to the increase of compound concentration along the trophic chain.

Biomarker of effect (effect biomarker)   biomarker that indicates an effect of a compound on the measured parameter.

Biomarker of exposure (exposure biomarker)   a measured parameter that indicates exposure to a compound. It need not be associated with any toxic effect.

Biomarker of susceptibility (susceptibility biomarker)a measured parameter that indicates that an organism is likely to be affected by a compound.

Biomonitoring   using organisms to evaluate the state of the environment (see Chapter 5).

Biota-sediment accumulation factor (BSAF) gives a quantitative value for the accumulation of a chemical in organisms and sediment.

Biotic ligand model (BLM) a model for estimating the influence of environmental factors on metal toxicity.

Biotransformation transformation of a compound by mechanisms in cells of an organism to daughter compounds.

BOD (biological oxygen demand) a measure that indicates the oxygen consumption of (mainly micro-) organisms in water.

Body burden the amount of a compound that an organism accumulates.

Carcinogenesis generation of cancer.

Carrier capacity (K) the maximal population size in a given space.

Caspases enzymes important in programmed cell death, apoptosis.

Catalase an antioxidant enzyme catalyzing the reaction from superoxide to water (see Table 11.3).

Chaperone a molecule assisting a protein in assuming or maintaining a correct structure when, for example, the structure would be destroyed by a toxicant in the absence of association to the chaperone molecule.

Chelating agents compounds that form a complex with a chemical (used especially in chemistry of metal ions).

Chloride cell gill cell type that actively regulates the ion balance of the animal. The cells are characterized by a very high number of mitochondria, enabling them to produce large amounts of ATP, which is used especially for active ion pumping by Na^+, K^+ ATPase, also present in high amounts in the cells. The difference in ion gradient between the freshwater and marine environments causes differences in the structure and properties of these cells between these environments.

Chlorosis bleaching of green plants because of a decreased amount of chlorophyll.

Chronic lethality lethality occurring as a result of prolonged exposure to a toxicant. The exposure is considered chronic if its length exceeds 10% of the life length of an organism (see Chapter 14).

Clastogenic (of a toxicant) capable of causing chromosomal damage in living cells.

Clearance in the context of ecotoxicology, the removal of a toxic compound from the compartment where it is accumulated.

COD (chemical oxygen demand) a parameter measured to evaluate water quality, indicating the oxygen consumption in water by abiotic components.

Comet assay (single-cell electrophoresis, usually alkaline single-cell electrophoresis). A method commonly used to indicate the presence of genotoxic agents that have caused DNA strand breakage. The name "comet assay" comes from the appearance of cellular DNA under the microscope after the electrophoresis. Undamaged DNA, which travels slowly, forms the head of the comet, whereas the tail is formed from the smaller, broken strand fragments. There are several programs estimating different aspects of damage to cellular DNA.

Common garden experiment a term used in ecological and evolutionary studies when all organisms used are reared in common environmental conditions throughout the experiment.

Community the different biotic components (different organism types) forming a system in a defined area.

Concentration addition (CA)   a model for similarly acting toxicants in which the toxicity depends solely on the concentrations (added together) of individual toxicants without interactions between the toxicants.

Congener   a member of a family of compounds with similar structures; the best known congeners are members of the PCB (polychlorinated biphenyl) family.

Conjugation   the most important reactions in phase 2 of detoxification. Conjugation comprises reactions with endogenous polar molecules (e.g. glucuronic acid, glutathione, and sulfate), usually to form organic ions that can be excreted in aqueous solution (see section 9.1.3).

Concentration-response   dependence of the response to a toxicant on its concentration.

Corticosteroids   hormones secreted from the cortical regions of adrenals/interrenals in mammals and hormones with similar structure in other vertebrates. Corticosteroids such as cortisol and cortisone are typically secreted under different stresses. For this reason, they are often called stress hormones.

Cross-resistance   exposure to a chemical causes increased tolerance/resistance to a different chemical.

Cytochrome P450   a group of heme-containing compounds important in detoxification (see Chapter 9).

Delta-aminolevulinic acid dehydratase (ALAD)   an important rate-limiting enzyme of heme synthesis. It is specifically inhibited by lead, making the enzyme a very good biomarker of lead intoxication.

Depuration   where an organism is allowed to get rid of the exposure chemical in a clean environment.

Developmental toxicity   disturbances caused by a toxicant in the developmental stages and rates of development.

Diffusion   passive movement of a compound from higher to lower concentration.

Direct effect   where a toxicant affects the organism directly.

DNA adduct   a toxicant attaches itself directly to the bases of DNA, most often guanosine. Consequently, the digestion product of DNA has higher molecular weight than the product of the same digestion in a non-exposed organism. The presence of DNA adducts is used as a biomarker showing the exposure to, for example, polyaromatic hydrocarbons.

DOC (dissolved organic carbon)   a measure of eutrophication of water; an increase in DOC is associated with eutrophication.

Dose   the amount of material reaching the organism (or its sensitive organs). In most cases, the dose is not the same as the ambient aquatic concentration, as various factors affect the uptake, metabolism, and excretion of a compound in an organism.

Dose–response   relates the amount of a chemical in the organism to its effects. If the amount of a compound in the organism per unit weight differs from what would be expected from the ambient aquatic concentration, concentration–response and dose–response are different.

Early-life-stage (ELS) test   a toxicity test using the stages from fertilization to early juveniles. The rationale for this type of testing is that usually the early life stages are more vulnerable to toxicants than adult animals, whereby virtually the same information about toxicity is obtained in a much shorter time than with full-life-cycle tests.

ECx   effective concentration of a toxicant for the measured endpoint. x indicates the percentage of organisms that show the effect.

EDx   effective dose of a toxicant. x indicates the percentage of organisms showing the effect.

Ecoregion   the abiotic components of ecosystems are similar, whereby they can be placed under one umbrella.

Ecosystem   comprises a space of a biological community and its abiotic environment.

Ecotoxicogenomics   the use of genomic methods (e.g. microarrays, quantitative polymerase chain reaction (PCR)) in studies of pollutants in an ecosystem.

Ecotoxicology   the ultimate goal is to understand toxicant effects on biological communities in an ecosystem. The practical definition is the study of toxicological effects on species other than man and laboratory rodents. However, even toxicological studies on humans may be ecotoxicological, if their aim is to show how the biological community is affected by toxicant effects on humans.

Effective population size   a population genetics concept that can be defined as the number of individuals in an idealized population that has the same value of any population genetic quantity as the population of interest.

Electronegativity   where a compound or some parts of its structure are negatively charged.

Elutriate   the wastewater entering the environment (e.g. from industry and water-treatment plants).

Endocrine disruptor   a chemical that affects the endocrine system at some point (see section 11.4).

End point   the measurement taken to estimate toxicological effect.

Endocytosis   uptake of a compound into a cell.

Enrichment factor   describes how effectively a compound is enriched in the measured compartment.

Environmental assessment   evaluation of how planned activity affects the environment.

EPA (US Environmental Protection Agency)   US government institution concerned with environmental affairs.

Epigenetic   transgenerational effects that do not involve a change in DNA structure.

Epizootic   a disease with much higher appearance in an animal population than is expected. High population densities, such as those occurring in aquaculture, increase the probability of epizootic diseases.

EROD (ethoxyresorufin-O-deethylase)   an enzyme whose activity is used to show changes in the activity of phase 1 of biotransformation, mainly by PAHs (polycyclic aromatic hydrocarbons) and other compounds that activate the aryl hydrocarbon receptor pathway.

Estrogenic chemicals   chemicals that are associated with female reproductive hormone pathways.

Euler–Lotka equation   allows an estimation of how a population is growing. The equation is:

$$1 = \sum_{a=1}^{\omega} \lambda^{-a} l(a) \, b(a),$$

where λ is the growth rate, $l(a)$ is the fraction of individuals reaching the age a, and $b(a)$ is the number of individuals born at time a.

Exocytosis   extrusion of material from a cell so that the compounds, enclosed in intracellular vesicles, are released to the extracellular compartment or environment. The vesicle membrane fuses with the cell membrane whereby the contents of the vesicle are released to the outside.

Facilitated diffusion   passive transport through membranes that occurs via specific membrane proteins. The transport proteins (transporters) can be quite selective for the compounds they accommodate. There are, for example, several different amino acid transporters. If the transporter exchanges compounds at different sides of the membrane, it is called an exchanger (e.g. anion exchanger). If several species are transported simultaneously in one direction, the transporter is called a cotransporter (e.g. potassium-chloride cotransporter). When the transporter uses an actively maintained (ion) gradient to transport compounds against their electrochemical gradient, and does not use energy stored in ATP directly, the transport is called secondarily active transport (e.g. sodium/proton exchange).

Feasibility study   preliminary experiment that indicates if the concept studied is worth pursuing further.

Fecundity   synonym of fertility.

Fenton reaction   a reaction of hydrogen peroxide in the presence of iron (and copper) that produces highly reactive hydroxyl radicals.

FETAX (frog embryo teratogenesis assay—*Xenopus*)   a test that evaluates the incidence of malformations in frog embryos caused by toxicants.

FMO (flavin-containing mono-oxygenase)   a group of enzymes in phase 1 of detoxification.

Fluctuating asymmetry   random asymmetry of organisms, which is thought to increase with contamination.

Free radical   an atom, molecule, or ion that has unpaired valence electrons or an open electron shell.

Fugacity   is close to the vapor pressure of a compound. The fugacity can be thought of as indicating the tendency of a compound to escape from the compartment it is presently attached to.

Gamma radiation   the type of radioactive radiation with the lowest energy but highest penetration.

Gavage   force-feeding.

Genetic bottleneck   where, as a result of a large decrease in the size of a population, the genetic variability forwarded to future generations is drastically reduced.

Genetic drift   a random change in the genetic variation in successive generations.

Genetic hitchhiking   where an allele is carried forward (to the next generation) because it is linked to another that is preferentially transferred.

Genetically modified organism (GMO)   an organism with man-made changes in its genome.

Genotoxicity   toxicity that appears in the genetic material.

Geographic information systems (GIS)   computerized systems for archiving, organizing, and analyzing spatial data.

Gill filament (primary lamellum)   the basic unit of gill that protrudes from gill arches and contains folds (secondary lamellae), which increase the respiratory area.

Global warming   a general warming of the earth that is thought to include an anthropogenic component. An increased level of carbon dioxide, which also causes ocean acidification, is thought to contribute to it.

Glucocorticoids   stress hormones produced by adrenal or interrenal cells of vertebrates, which affect glucose balance (and immune function). See corticosteroids.

Glutathione   the major small molecule regulating the redox balance (a redox buffer). Glutathione is a tripeptide of cysteine, glutamate, and glycine. The sulfhydryl group of the cysteine is the major player in the redox reactions.

Glutathione-S-transferase (GST)   A phase 2 conjugating enzyme (see section 9.1.3).

Haber–Weiss reaction   the reaction where a hydroxyl radical is formed from superoxide via hydrogen peroxide.

Hardy–Weinberg equilibrium the genetic composition of a population remains constant if it is large, not subject to selection, and the mutation rate and migration are negligible.

Hazard the potential of a chemical to cause harm.

Hazard assessment comparison of the potential of chemical to cause harm with its concentration in the environment.

Hazard quotient (HQ) an indicator of hazard expected on the basis of expected environmental concentration (EEC) and estimated threshold effect concentration (EFE). HQ = EEC/EFE.

Heat shock protein (HSP) stress proteins of different molecular weight classes. The proteins are molecular chaperones. Some are constitutively formed, but many are induced by stresses affecting protein conformation. Many HSPs help in refolding to original conformation, but some assist in protein breakdown.

Henderson–Hasselbalch equation the major acid–base equilibrium equation relating pH, the acid equilibrium constant (K_a), and concentrations of the acid and base forms of the conjugate acid. $pH = pK_a + \log([B^-]/[BH])$.

Henry's law the concentration of a compound in the aqueous phase is proportional to its partial pressure. (Depending additionally on the water solubility of the compound.)

Heterozygosity the degree of genetic variation (variation of alleles).

Histidine imidazole the imidazole group of histidines is the major pH-sensitive group of proteins in the physiologically important pH range (6–8). The protonation of the imidazole group changes with pH and can also be affected by several contaminants.

Hormesis the concentration-response relationships of influx and efflux are different. Influx in low concentration is stimulated and efflux is inhibited.

Hypoxia where the oxygen level in water is clearly below atmospheric. Whenever one is talking of hypoxia, the degree of hypoxia (e.g. % of air saturation) should be given, as the tolerance of low oxygen levels varies.

Hyperoxia where the oxygen level in water is above atmospheric. Hyperoxic conditions may occur in eutrophic areas in the daytime when green plants produce oxygen during active photosynthesis.

Hyperplasia an increase in the number of cells in a tissue.

Hypertrophy an increase in the size of cells in a tissue.

Imposex the presence of reversed sexual characters as a response to toxicants. The term has been used especially for the appearance of male sexual characters (penis) in females of prosobranch molluscs as a response to organic tin exposure.

Independent action (IA) where toxicants have different sites and modes of action.

Indirect effect effect of a toxicant that is caused by an effect on another species, not a direct chemical effect on the species itself. As an example, if the availability of shelter is affected by a herbicide effect on aquatic plants, the population of animals may be affected even if the chemical has no effect on the animals.

Ischemia drastically reduced blood supply to a tissue, leading to inadequate oxygen availability.

Isoform when there are several forms of a protein, the individual types are called isoforms.

Isozymes if different types of an enzyme exist, the different forms are called isozymes.

Keystone species a species that has a large influence on the community that it is part of (the effect is much larger than the abundance of the species would suggest).

K_{ow} octanol–water partition coefficient. The value is used as a measure of the lipid solubility of a chemical.

LCx   lethal concentration for x% of organisms. The value is highly dependent on the time of measurement, which thus needs to be given.

LDx   lethal dose for x% of organisms. The value is highly dependent on the time of measurement, which thus needs to be given.

Lesion   an abnormality in (damage to) a tissue of an organism, usually caused by toxicants.

Life-cycle test   toxicity testing from reproduction of parents to the first reproduction of offspring.

Lipid peroxidation (LPO)   oxidative effect observed as oxidation of lipid moieties. Often measured as thiobarbituric-acid-reactive substances (TBARS).

Lipid solubility   the solubility of a compound in lipid phase, usually determined with octanol–water partitioning. An increase in K_{ow} indicates an increase in lipid solubility.

LOEC (lowest observed effect concentration)   the lowest concentration of a toxicant that causes a measurable effect. The effect is often estimated from probit analysis.

Logit   an inverse of sigmoidal logistic function, occasionally used to analyze data from toxicity testing.

Malonyl dialdehyde (MDA)   a major end product of oxidative damage to lipids, lipid peroxidation.

Mean generation time   the average time between two consecutive generations.

Mean residence time   in toxicology, the amount of time that the contaminant stays in the compartment (aquatic phase, sediment).

Mesocosm   experimental unit with several trophic levels (see Chapter 5). The use of mesocosms tries to combine the reproducibly of laboratory experiments with environmental realism. To some degree, enables estimating the indirect effects of toxicants.

Metabolic scope   gives the difference between minimum (resting) and maximal (active) metabolism. Many toxicants affect this difference.

Metallothionein   sulfhydryl-group-rich protein(s), which bind metal ions. The binding affinity is metal-dependent. In particular, complexation of cadmium, copper, and zinc has been studied.

Metapopulations   virtually independent units of a population that is divided to fragments (each fragment is a metapopulation).

Methemoglobin   the form of hemoglobin in which the heme iron has been oxidized to the ferric (Fe^{3+}) state and is unable to bind oxygen. Normal hemoglobin is changed to methemoglobin by oxidative stresses—e.g. exposure to nitrite.

Microcosm   a defined experimental unit containing several trophic levels. Smaller and simpler than a mesocosm.

Micronucleus   a part of the genomic material that is not enclosed in the nucleus proper, but forms a smaller entity. A micronucleus is defined as being maximally 1/3 of the size of the nucleus proper, and separate from it. In addition to micronuclei, nuclear abnormalities may occur as lobed nuclei or nuclei with irregular shape. All of these indicate gross DNA damage resulting in unnatural movement of chromosomes in cell division.

Microtox assay   Bacteria (*Vibrio fischeri*) that luminesce when exposed to a toxicant are acutely exposed to aqueous samples. The light output given is compared to the light output without the sample.

Mixed function oxidases (monooxygenases)   a group of phase 1 detoxification enzymes (see section 9.1.2).

MOA (mode of action)   the way by which a toxicant affects an organism.

MTF-1 (metal transcription factor-1)   a transcription factor that binds to the metal response element in the promoter/enhancer region of a gene when Zn^{2+} has bound to it.

Mutagen   a compound that causes an increased mutation rate.

MXR (multixenobiotic resistance)   the mechanism by which invertebrates, especially shellfish, remove organic toxicants. ABC transporters are an important component of MXR.

Nanomaterial   a material which is less than 10 nm thick, and less than 100 nm long and wide. Nano-materials are often metal (e.g. titanium, ferric, silver) oxides associated with carbon (e.g. fullerene). They are increasingly used in many applications, such as waterproofing textiles.

Neoplasia   abnormal tissue formation (tumor) as a result of abnormal growth or division of cells.

Niche   the (unique) position of a species in a community (spatial, trophic level, etc.).

NOEC ((highest) no observed effect concentration)   the highest toxicant concentration that does not cause an effect.

Nontarget organism   an organism experiencing the effect of, for example, a pesticide, although the pesticide has not been directed to it.

NOTEL (no observed transcriptional effect level)   the highest toxicant concentration that does not cause a change in gene transcription.

Ocean acidification   decrease of the pH of the oceans, which is attributed to an increase in carbon dioxide production or its removal from oceanic water.

Oxidative stress   a disturbance of the redox balance to the oxidizing direction. Oxidative stresses are associated with increased formation of reactive oxygen species (ROS) (see section 11.3).

Oxyradical   all oxygen radicals are ROS, but not all ROS are oxyradicals. Oxyradicals are oxygen species with unpaired electrons or accessible electron shells.

QSAR (quantitative structure–activity relationship)   indicates cases (especially of some organic molecules) where the probable activity (toxic action) can be estimated based on the molecular structure of the compound.

qPCR (quantitative polymerase chain reaction)   a commonly used genomic method that quantifies the amount of transcript in a sample. The most commonly used quantification is relative, requiring that the mRNA production of a gene, to which the transcription of the evaluated gene is compared, remains constant throughout the experiment.

PAHs (polycyclic aromatic hydrocarbons)   a group of organic toxicants.

Periphyton   a complex mixture of (largely autotrophic) microbes on the surfaces of submerged structures of water bodies. It serves as food for zooplankton and invertebrates generally.

Peroxisome   an organelle of eukaryotic cells involved in the catabolism of long-chained and branched fatty acids and polyamines. Various biosynthetic functions have also been suggested for them.

Persistent organic pollutants (POP)   very stable organic compounds that exert their toxicity for years after they have entered the environment. PCBs (polychlorinated biphenyls) and DDT-type insecticides are typical examples of such compounds.

Pharmaceuticals and personal care products   an emerging group of pollutants including drugs, soaps, deodorants, sunscreens, etc. Includes many different types of organic compounds.

Pharmacokinetics   the absorption, distribution, metabolism, and excretion of pharmaceuticals.

Phase 1   the first part of organic compound detoxification (see section 9.1.2).

Phase 2   the second part of organic compound detoxification (see section 9.1.3).

Phenotypic plasticity   the possibility of phenotypes of organisms to react to environmental factors (also called individual or physiological adaptation).

Photosensitivity   the degree to which an organism (or its tissues) is sensitive to light.

Phytochelatins   metal-binding compounds found especially in plants, but recently also found in several invertebrates.

Potentiation   when toxicants are considered, the toxic effect of a compound is increased in the presence of a second compound.

Probit   probit analysis is the most common analysis method in toxicological studies (toxicity tests). It is a type of regression where the dependent value can have only two values (dead or alive, affected or not affected).

Range-finding tests   preliminary studies that aim to find the effective concentration range of a toxicant, which is then used in the actual detailed studies.

Rare earth element   fifteen lanthanides, scandium, and yttrium. The elements are mainly mined in China and used especially in various information technology products.

REACH (Registration, Evaluation, Authorisation and Restriction of Chemicals)   The chemical regulation initiative of European Community.

Risk assessment   in aquatic toxicology, evaluating the risk caused by contamination to the organisms studied (see section 18.2).

Scope of activity   the difference between resting and maximal metabolism. The term is virtually synonymous with metabolic scope.

Scope for growth   the amount of energy available for growth.

Secondary lamellae   the respiratory surfaces of gills in fish. Gills consist of gill arches, from which the lamellae point outward. The lamellae are folded, and folds are called secondary lamellae.

Sentinel species   indicator organisms used in biomonitoring to assess contaminant effects.

Shannon diversity index   estimation of biological diversity based on the geometric mean of the diversity.

Single-cell electrophoresis   usually done as the comet assay (which is most often alkaline single-cell electrophoresis) to show DNA strand breakage. The comet head (with very slow electrophoretic movement of the high-molecular-weight DNA) shows unbroken DNA strands, whereas the comet tails (where DNA moves faster in the electric gradient) indicate the presence of DNA fragments.

Sorption   a process by which compounds become attached to each other as, for example, when a compound is attached to sediment.

Species assembly   the group of species making up a community.

Species diversity   the number of species that is found in a collection of individuals (from a community).

Specific activity   normally, radioactivity in a given weight of a compound. The term can also be used for other types of activity when they are related to a unit amount of a substance.

Spiking   addition of chemical to the experimental medium (spiking is used especially for sediment toxicity).

Spillover   where an amount of toxicant enters the environment outside of the area/container where it is intended to remain.

SSH (subtractive suppression hybridization)   a method for treating microarray samples to decrease the amount of redundant sequences that need to be analyzed. SSH produces cDNA (complementary DNA) libraries that contain mainly the transcripts that are affected by the exposure.

Stable isotope assay   in ecology, a method used to evaluate the food components of an organism. Every food type has a unique composition of stable isotopes, so that their analysis enables evaluation of the basic nutrition type.

Static-renewal toxicity test   where the medium used in the test is replaced at set intervals.

Static toxicity test   toxicity test in which the medium and organisms are added to the test and not replaced during the test.

Stress   any change in the biotic or abiotic environment of the organism that causes specific reactions in the organism.

Stressor   the change that is behind the stress reactions of an organism.

Stress protein   see heat shock protein (HSP), which these proteins were initially called and often still are.

Sublethal effect   an effect observed that occurs without the death of the organism.

Sulfotransferase (SULT)   Phase 2 conjugating enzyme group (see section 9.1.3).

Synergism   where different chemicals strengthen each other's actions.

Target organ   an organ in which the effect of a chemical is expected to be found.

Target organism   the species or group of species (e.g. insects) towards which the toxic action of a chemical is intended.

Teratogen   a compound that causes the formation of abnormal structures.

TOC (total organic carbon)   the total amount of carbon in a sample (e.g. water sample, sediment sample).

Toxic equivalency (TEQ)   the toxicity of a compound relative to a standard compound. The most common use of toxic equivalency is to relate the toxicity of PCBs (polychlorinated biphenyls) to a dioxin compound (TCDD).

Toxic equivalency factor (TEF)   when many different congeners are present in a mixture of compounds, their toxicity is combined to a single number using their relative toxicity in comparison to a standard compound and their concentration; calculated as:

$$TEF = C_a TEQ_a + C_b TEQ_b + ... + C_n TEQ_n,$$

where C is the concentration of a compound (a to n).

Toxin   a chemical from the natural environment that causes adverse effects in organisms; for example, cyanobacterial toxins.

Toxicant   a chemical which has adverse effects on the function of organisms; can be either natural or man-made.

Trophic level   describes the level of an organism in the food chain (e.g. phytoplankton, planktivorous fish).

Trophic transfer   transfer of a compound from one level in the food chain to the next (e.g. from zooplankton to the fish eating them).

UDP-glucuronosyltransferase   Phase 2 conjugating enzyme (see section 9.1.3).

Uncertainty factors   arbitrarily decided factors that try to take into account the uncertainty of a chemical causing organismic effect (and thus acting as a toxicant). Commonly used uncertainty factors are 10, 100, and 1000. The accepted environmental concentration of a compound is the concentration known to cause an effect divided by the uncertainty factor. The better known a compound's effects are, the smaller the uncertainty factor.

Vitellogenin   precursor of yolk proteins. In vertebrates, mainly produced in the liver. Cleaved to the smaller yolk constituents.

Xenobiotic   a compound foreign to the environment.

Xenoestrogen   a compound that mimics estrogen function, but is not a natural feature of the estrogen cycle.

Index

Note: Page numbers followed by f indicate figures; t, tables.